Climate Change, Climatic Extremes, and Human Societies in the Past

Climate Change, Climatic Extremes, and Human Societies in the Past

Editor

Harry F. Lee

MDPI • Basel • Beijing • Wuhan • Barcelona • Belgrade • Manchester • Tokyo • Cluj • Tianjin

Editor
Harry F. Lee
Department of Geography and
Resource Management,
The Chinese University of Hong Kong
China

Editorial Office
MDPI
St. Alban-Anlage 66
4052 Basel, Switzerland

This is a reprint of articles from the Special Issue published online in the open access journal *Atmosphere* (ISSN 2073-4433) (available at: https://www.mdpi.com/journal/atmosphere/special_issues/climate_change_extremes).

For citation purposes, cite each article independently as indicated on the article page online and as indicated below:

LastName, A.A.; LastName, B.B.; LastName, C.C. Article Title. *Journal Name* **Year**, *Article Number*, Page Range.

ISBN 978-3-03936-960-7 (Hbk)
ISBN 978-3-03936-961-4 (PDF)

Contents

About the Editor

Harry F. Lee graduated from The University of Hong Kong. He is now the Associate Professor of the Department of Geography and Resource Management, The Chinese University of Hong Kong. He is interested in the study on socio-economic and demographic impacts of climate change in ancient and recent human history, historical epidemiology, underlying mechanisms of climatic extremes, and environmental perceptions and sustainability.

Editorial

Climate Change, Climatic Extremes, and Human Societies in the Past

Harry F. Lee

Department of Geography and Resource Management, The Chinese University of Hong Kong, Shatin, New Territories, Hong Kong, China; harrylee@cuhk.edu.hk

Received: 7 July 2020; Accepted: 17 July 2020; Published: 20 July 2020

1. Introduction

More people appreciate the importance of global sustainability. This is evidenced by a growing number of quantitative studies investigating the connection between climate change and human societies. Given this background, the *Atmosphere* Special Issue "Climate Change, Climatic Extremes, and Human Societies in the Past" aims to highlight the major aspects of the climate–society nexus in ancient and recent human history. There are eight papers in this Special Issue based on quantitative approaches which illustrate different forms of the climate–society nexus in ancient (two papers), historical (three papers), and contemporary periods (three papers).

2. Temporal Coverage of this Special Issue

2.1. Ancient Periods

Regarding ancient periods, the interconnection among climate, agriculture, and human societies is assessed. Li et al. [1] review archaeobotanical evidence from Neolithic sites in China and show that rice was primarily cultivated in the Yangtze River valley and its southern edge, while millet cultivation occurred in northern China circa 9000–7000 BP. Millet- and rice-based agriculture intensified and expanded during 7000–5000 BP. In 5000–4000 BP, rice agriculture continued to develop in the Yangtze River valley, and millet cultivation moved gradually westwards. Meanwhile, mixed agriculture based on both millet and rice developed along the border between the North and South. Climate-induced changes in vegetation and the environment played a significant role in agricultural development from 7000–6000 BP, while precipitation was crucial in shaping the distinct regional patterns of Chinese agriculture from 6000–4000 BP.

While climate and agriculture were closely connected in ancient times, the social dynamics in human societies were also thought important, significantly mediating the climate–agriculture connection. Wang et al. [2] base their paper on the human bone fragments obtained from the site of Xiaohucun, dated to the late Shang Dynasty (ca. 1250–1046 BC) in China, together with the isotopic analysis of collagen, to illustrate the connections between social status and diet. Those elite members probably consumed more animal protein such as horses, pigs, donkeys, and sheep/goats than the common people in the late Shang Dynasty.

2.2. Historical Periods

Regarding historical periods, various positive checks such as wars, famines, and epidemics are examined in this Special Issue. The common theme of the associated papers is to reveal the non-linear and complex relationship between climate change and the positive checks in historical China and pre-industrial Europe. Zhang et al. [3] employ Emerging Hot Spot Analysis to examine war hot spots in China from 1–1911. They show that war hot spots were generally located in the Loess Plateau and the North China Plain during warm and wet periods, but in the Central Plain, the Jianghuai area, and the lower reaches of the Yangtze River during cold and dry periods. Furthermore, the hot spots for agri-nomadic warfare had the abovementioned trends, while rebellion hot spots expanded outward during warm and wet phases and compressed inward during cold and dry phases.

Zhai et al. [4] investigate the social responses to the North China Famine of 1876–1879, which was brought on by extreme drought. They show that famine-related migration tended to be spontaneous and short-distanced, with the flow mainly spreading to the surrounding areas and towns. Furthermore, relief-money and grain from the non-disaster areas were allocated to the disaster areas. Yet, such state administrated intervention disrupted the equilibrium of food markets in non-disaster regions, resulting in food price fluctuations there.

Yue and Lee [5] examine the relative impact of the direct and indirect impacts of climate change on plague outbreaks in Europe between 1347–1760 using Structural Equation Models. They found that all of the climatic impacts on plague outbreaks were indirect and were materialized through economic changes. They further demonstrated that temperature-induced economic changes triggered plague outbreaks in cold and wet periods, while precipitation-induced economic changes induced plague outbreaks in cold periods.

2.3. Contemporary Periods

Over more recent times, the papers in this Special Issue focus on weather-related phenomena which significantly affect human societies. The non-linear dynamics of those phenomena are also highlighted. The associated findings can help human societies to mitigate the adverse impacts of weather extremes better. Xiang et al. [6] base their paper on summer precipitation data and 130 circulation indexes of 34 national meteorological stations in Chongqing spanning 1961–2010, together with the decision tree and the stochastic forest algorithm, to build a new multi-factor model for summer precipitation in Chongqing. Moreover, the model is tested with precipitation data from 2011–2018. Results show that the new model outperforms previous single-factor models.

Zhou et al. [7] use temperature data over 1980–2012 together with the Correlation Dimension method to analyze the temperature dynamics in the Yangtze River Delta in China. They find that the temperature rose by 1.53 °C over this period and the temperature rose the fastest in densely populated urban areas. However, the temperature dynamics were more complicated in the sparsely populated areas when compared to densely populated urban areas. Moreover, the complexity of temperature dynamics increased along with the increase in temporal scale. Lastly, the interaction between economic activity and urban density had the most substantial influence on temperature.

Yuan et al. [8] investigate the coupling between soil moisture and air temperature over China spanning from 1980–2013 using an energy-based diagnostic process. They show that the soil moisture–temperature coupling is the highest in the transition zones between wet and dry climates (e.g., north-eastern China and part of the Tibetan Plateau). Furthermore, the coupling is stronger in spring, and varies greatly in different seasons over different climatic zones. The heatwaves of 2009 in North China and 2013 in Southeast China further reveal that regions having low soil moisture may enhance heat anomalies, which further strengthens the coupling between soil moisture and temperature.

3. Conclusions

In summary, this Special Issue contributes theoretical and methodological analyses of the climate—society nexus. However, the conceptualization of the climate–society nexus is not a binary one. The nexus should be contextualized, while interdisciplinary collaboration should be further sought for addressing the topic [9,10]. It is also worth noting that the climate–society nexus is also dependent on temporal and spatial scales, and research findings will be determined by the length of the study time span and the size of the study area [11–13]. The aim of this Special Issue is to facilitate a more fruitful discussion about the climate–society nexus.

I am very thankful to my colleagues for their invaluable contributions and the reviewers for constructive comments and suggestions that helped to improve the papers. Furthermore, I thank the MDPI journal office for their excellent support in processing and publishing this Special Issue.

Funding: This study is supported by the Improvement on Competitiveness in Hiring New Faculties Funding Scheme (4930900) and Direct Grant for Research 2018/19 (4052199) of the Chinese University of Hong Kong.

Conflicts of Interest: The author declare no conflict of interest.

References

1. Li, R.; Lv, F.; Yang, L.; Liu, F.; Liu, R.; Dong, G. Spatial-temporal variation of cropping patterns in relation to climate change in Neolithic China. *Atmosphere* **2020**, *11*, 677. [CrossRef]
2. Wang, N.; Jia, L.; Si, Y.; Jia, X. Isotopic results reveal possible links between diet and social status in Late Shang Dynasty (ca. 1250–1046 BC) tombs at Xiaohucun, China. *Atmosphere* **2020**, *11*, 451.
3. Zhang, S.; Zhang, D.D.; Li, J. Climate change and the pattern of the hot spots of war in ancient China. *Atmosphere* **2020**, *11*, 378. [CrossRef]
4. Zhai, X.; Fang, X.; Su, Y. Regional interactions in social responses to extreme climate events: A case study of the North China Famine of 1876–1879. *Atmosphere* **2020**, *11*, 393. [CrossRef]
5. Yue, R.P.H.; Lee, H.F. Examining the direct and indirect effects of climatic variables on plague dynamics. *Atmosphere* **2020**, *11*, 388. [CrossRef]
6. Xiang, B.; Zeng, C.; Dong, X.; Wang, J. The application of a decision tree and stochastic forest model in summer precipitation prediction in Chongqing. *Atmosphere* **2020**, *11*, 508. [CrossRef]
7. Zhou, C.; Zhu, N.; Xu, J.; Yang, D. The contribution rate of driving factors and their interactions to temperature in the Yangtze River Delta Region. *Atmosphere* **2020**, *11*, 32. [CrossRef]
8. Yuan, Q.; Wang, G.; Zhu, C.; Lou, D.; Hagan, D.F.T.; Ma, X.; Zhan, M. Coupling of soil moisture and air temperature from multiyear data during 1980–2013 over China. *Atmosphere* **2020**, *11*, 25. [CrossRef]
9. Lee, H.F. Historical climate-war nexus in the eyes of geographers. *Asian Geogr.* 2020. [CrossRef]
10. Lee, H.F. Measuring the effect of climate change on wars in history. *Asian Geogr.* **2018**, *35*, 123–142. [CrossRef]
11. Dong, G.; Liu, F.; Chen, F. Environmental and technological effects on ancient social evolution at different spatial scales. *Sci. China Earth Sci.* **2017**, *60*, 2067–2077. [CrossRef]
12. Lee, H.F.; Fei, J.; Chan, C.Y.S.; Pei, Q.; Jia, X.; Yue, R.P.H. Climate change and epidemics in Chinese history: A multi-scalar analysis. *Soc. Sci. Med.* **2017**, *174*, 53–63. [CrossRef] [PubMed]
13. Lee, H.F. Cannibalism in northern China between 1470 and 1911. *Reg. Environ. Chang.* **2019**, *19*, 2573–2581. [CrossRef]

Article

Spatial–Temporal Variation of Cropping Patterns in Relation to Climate Change in Neolithic China

Ruo Li [1], Feiya Lv [1,2], Liu Yang [1], Fengwen Liu [1,3], Ruiliang Liu [4,*] and Guanghui Dong [1,5,*]

[1]   Key Laboratory of Western China's Environmental System (Ministry of Education),
     College of Earth and Environmental Sciences, Lanzhou University, Lanzhou 730000, China;
     lir2014@lzu.edu.cn (R.L.); lvfy14@lzu.edu.cn (F.L.); lyang2018@lzu.edu.cn (L.Y.); liufw@ynu.edu.cn (F.L.)
[2]   Department of Geography, University of Wisconsin, Madison, WI 53706, USA
[3]   School of Resource Environment and Earth Science, Yunnan University, Kunming 650000, China
[4]   School of Archaeology, University of Oxford, Oxford OX1 3TG, UK
[5]   CAS Center for Excellence in Tibetan Plateau Earth Sciences, Chinese Academy of Sciences (CAS),
     Beijing 100101, China
*   Correspondence: ruiliang.liu@arch.ox.ac.uk (R.L.); ghdong@lzu.edu.cn (G.D.)

Received: 14 May 2020; Accepted: 19 June 2020; Published: 27 June 2020

Abstract: The Neolithic period witnessed the start and spread of agriculture across Eurasia, as well as the beginning of important climate changes which would take place over millennia. Nevertheless, it remains rather unclear in what ways local societies chose to respond to these considerable changes in both the shorter and longer term. Crops such as rice and millet were domesticated in the Yangtze River and the Yellow River valleys in China during the early Holocene. Paleoclimate studies suggest that the pattern of precipitation in these two areas was distinctly different. This paper reviews updated archaeobotanical evidence from Neolithic sites in China. Comparing these results to the regional high-resolution paleoclimate records enables us to better understand the development of rice and millet and its relation to climate change. This comparison shows that rice was mainly cultivated in the Yangtze River valley and its southern margin, whereas millet cultivation occurred in the northern area of China during 9000–7000 BP. Both millet and rice-based agriculture became intensified and expanded during 7000–5000 BP. In the following period of 5000–4000 BP, rice agriculture continued to expand within the Yangtze River valley and millet cultivation moved gradually westwards. Meanwhile, mixed agriculture based on both millet and rice developed along the boundary between north and south. From 9000–7000 BP, China maintained hunting activities. Subsequently, from 7000–6000 BP, changes in vegetation and landscape triggered by climate change played an essential role in the development of agriculture. Precipitation became an important factor in forming the distinct regional patterns of Chinese agriculture in 6000–4000 BP.

Keywords: Yangtze River valley; Yellow River valley; rice cultivation; millet cultivation; precipitation; Neolithic China

---

1. Introduction

The climate is one of the driving forces behind the social evolution of humans, especially in prehistory [1–4]. The studies of different strategies adopted by human societies in response to drastic climate fluctuation in the past can provide valuable insights into the underlying patterns and mechanisms of the human–land relationship. They can also offer important lessons on coping with the current challenges of rapid climate change in the modern world, such as global warming. The Neolithic period coincided with the early-mid Holocene—a recent warming period with numerous considerable climate fluctuations. It is one of the most significant stages of sociocultural evolution in human history. In recent years, there has been an increase of research focusing on human–environment interaction in this specific period [4–8].

One of the era-defining events in the Neolithic was the development of agriculture across the old world, followed by a substantial increase in the size of population and settlements [9–12]. Climate change has been considered to be a critical factor in the emergence and intensification of agriculture during the Neolithic period [13,14]. While the changes in temperature follow the same approximate trends in different regions of the Northern Hemisphere [15], precipitation shows distinct patterns affected by local climate (e.g., the arid central Asia and the Asian Monsoon Region) [16,17].

East Asia was one of the points of origin of agricultural development. Rice and broomcorn/foxtail millet were domesticated around 10,000 BP in the Yangtze River valley (southern China) and the Yellow River valley (northern China), respectively [18–20]. Both crops later became widely cultivated and formed the well-known agricultural structure in Neolithic China of northern millet vs southern rice [21]. Paleoclimate studies also suggest a similar geographical distinction between the Yangtze and the Yellow River valley in terms of moisture variation, called the anti-phase pattern [22]. The relationship between agricultural development in China and the regional variation of precipitation has not yet been discussed in detail. In this paper, we have collected and analyzed a large amount of legacy data, containing both archaeobotanical remains and radiocarbon dates from 125 Neolithic sites in China. Correlating the archaeobotanical evidence with the variation of precipitation reconstructed from well-dated paleoclimate records fills a significant gap in the current literature and allows us re-explore the different phases of Neolithic China in more detail.

2. Spatial–Temporal Change of Human Cropping Structures in Neolithic China

It has been widely agreed that the Neolithic cultural evolution in northern China can be divided into three phases: the pre-Yangshao period (9000–7000 BP), the Yangshao period (7000–5000 BP) and the Longshan period (5000–4000 BP). This chronological framework is based on multiple lines of evidence ranging from material culture (e.g., pottery typology), settlement, dietary practice, technological change, stratigraphy and radiocarbon dating [23–25]. It offers a crucial point of departure for us to reconstruct spatial patterns of the development of agriculture in China based on the archaeobotanical evidence, which in return contributes to the overall picture. We collected archaeobotanical data from 125 Neolithic sites across China: the data recorded include the presence and absence of targeted crops (millet, rice, barley, wheat), their absolute quantities and ratios between different types of crops. Soybean crops have been excluded from this work, although carbonized grains have occasionally been identified at some Neolithic sites in China, owing to the difficulty in separating domesticated from wild varieties.

As is the case with legacy collections, the quality of data varies considerably. The first level of complexity is caused by the scale of the archaeological excavation, as a fully excavated site might present a more complete picture of crops in use at a given time in comparison to samples collected from a surface survey. The data may have also derived from different theoretical and methodological approaches prioritizing certain practices or scrutiny above others, and finally the identification of crop remains is often subject to the individual analyst's experience (e.g., wild crops vs domesticated crops). Despite these potential complexities, the broad changes indicated by our big-data approach are arguably valid and important, as the broad patterns are not based on any specific site or site type and there are always multiple sites within one region for cross-checking. Furthermore, additional data, such as stable isotopes, are also included in the discussion and help to counter-check the archaeological narratives.

3. Spatial Pattern of Cropping Structures in China during 9000–7000 BP

Archaeobotanical studies suggest that the initial domestication of rice took place in the middle reaches of the Yangtze River prior to 10,000 BP, whereas broomcorn and foxtail millet were first domesticated in the Central Plains of northern China around 10,000 BP [18–20,26]. However, due to the limited number of identifiable crop fossils in these early Neolithic sites and the lack of direct radiocarbon dating, the exact chronology of the earliest millet/rice remains highly controversial. As macrofossils of domesticated millet and rice become more ubiquitous in sites dated between 9000 and 7000 BP and have direct radiocarbon dates associated with them, it is possible to say that the cultivation of rice and millet dated at least as far back as 9000 BP [19,27–30]. Macro and micro fossils of rice dated to 9000–7000 BP were unearthed from sites in the middle and lower reaches of the Yangtze River, such as Shangshan, Xiaohuangshan and Pengtoushan [20]. These sites are mainly in the piedmont zones, possibly owing to locational convenience for hunting and gathering activities, which were the primary survival strategies during this period. Interestingly, rice appears also to be the most ubiquitous crop in a few sites in the southern margin of the middle Yellow River valley (Figure 1), as exemplified by the famous Jiahu site, one of the largest settlements of the pre-Yangshao culture in the present-day Henan Province [31,32].

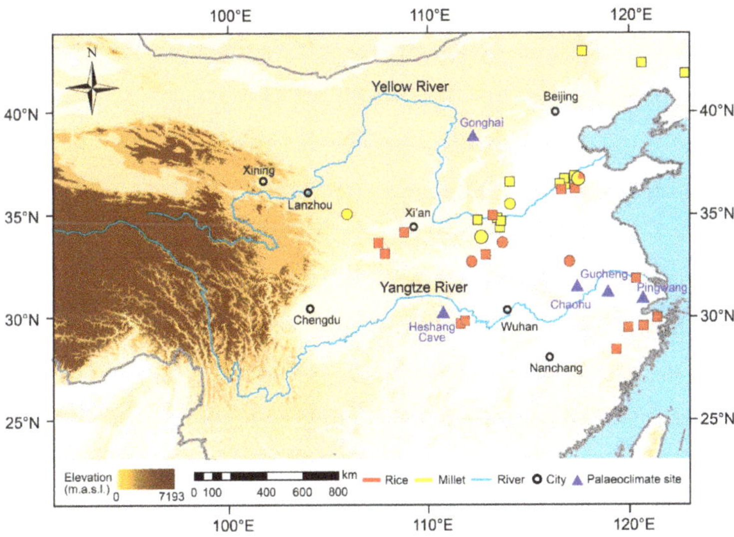

Figure 1. The spatial pattern of crop macro-fossils from sites dated between 9000–7000 BP in China. Squares represent sites without detailed archaeobotanical data; circles represent sites with detailed archaeobotanical data.

Charred grains of broomcorn and foxtail millet have been identified from numerous pre-Yangshao cultures (9000–7000 BP) in northern China, including the Peiligang, Xinglongwa, Houli, Cishan and Laoguantai-Dadiwan [19,27,28,33]. In some sites of the Peiligang and Houli cultures, both millet and rice were cultivated together (Figure 1). Most pre-Yangshao sites in northern China were located in the foothill areas, probably to facilitate hunting and gathering [34]. Stable carbon isotopes of human bones show a clear C_3 signal [35,36], indicating that hunting and gathering—rather than millet cultivation—was the major food strategy in this period [37]. Based on the large number of charred grains of foxtail and broomcorn millet as well as the clear C_4 signal from the isotope analysis of human bones in the sites of Xinglonggou and Xinglongwa in east Inner Mongolia [38], it is possible that millet cultivation might have become the principal food strategy in northern China during 8000–7000 BP [39].

4. Spatial Pattern of Human Cropping Structures in China during 7000–5000 BP

Evidence from the Yangshao Period sites in China (7000–5000 BP) suggests that, by this time, crop cultivation had replaced hunting/gathering and was the primary subsistence strategy in both the Yangtze and Yellow River valleys; for example, huge amounts of rice remains have been found in the storage pits of the Hemudu site [40]. Moreover, the ratio of domesticated rice to wild rice increased rapidly between ~6700–6300 BP in the lower Yangtze River valley [41]. In the Yellow River valley, carbon isotopes of human bones unearthed from almost all Yangshao sites display a clear C_4 signal.

This strongly indicates that millet became a routine part of diet in northern China between 7000 and 5000 BP [39].

The geographical distribution of both millet and rice from 7000 to 5000 BP is undoubtedly larger than it was from 9000 to 7000 BP (Figures 1 and 2). During the Yangshao period (7000–5000 BP), the regular practice of rice cultivation extended further northwestward to the west Loess Plateau; one example of this is the presence of domesticated rice at Xishanping, a site in the western Loess Plateau [42]. The distribution of millet expanded in several directions: eastward to the Shandong Peninsula and even Korea [43], westward to the northeast Tibetan Plateau [44], southwestward to the Chengdu Plain [45] and southward to the middle Yangtze River valley (Figure 2; [46]).

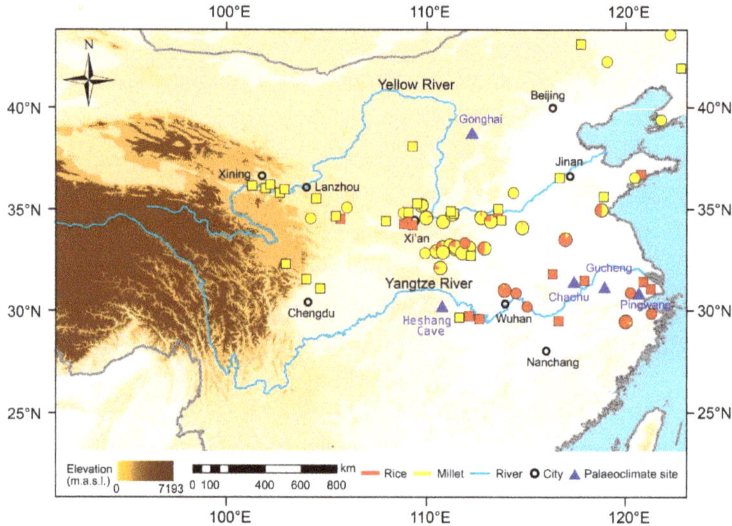

Figure 2. The spatial pattern of crop macro-fossils from sites dated between 7000–5000 BP in China. Squares represent sites without detailed archaeobotanical data; circles represent sites with detailed archaeobotanical data.

5. Spatial Pattern of Human Cropping Structures in China during 5000–4000 BP

After the Yangshao period, there was a broad westward movement of millet farming in the Longshan period (5000–4000 BP, Figures 2 and 3). In the east costal area of Shandong, local people replaced millet with rice during 5000–4000 BP. In contrast, farmers of the Machang culture in western China (4300–4000 BP) continued to rely heavily on millet cultivation. Together with the movement of farmers, millet moved gradually westward and was cultivated in the Hexi Corridor [47] and east Xinjiang [48]. Foxtail millet has also been identified at Karuo, a site on the southeastern Tibetan Plateau, which has been dated to ~4700–4300 BP [49]. One should bear in mind that these crops were also likely to be exchanged from adjacent farming societies rather than being locally cultivated [50].

The Yangtze River valley saw an expansion of rice cultivation during the Longshan period (5000–4000 BP) when rice was undoubtedly the dominant crop in the region. During the same period, it replaced millet and became the major subsistence crop in the northwest Chengdu basin as well (Figure 3; [45]). Meanwhile, rice cultivation spread westward to the Yunnan-Guizhou Plateau [51], as exemplified by the large number of rice and millet macrofossils from the first phase of the Baiyangcun site in northwest Yunnan [52]. It was further introduced into the Pearl River Delta of southern China during 5000–4000 BP. Charred rice grains have been discovered in both Laoyuan and Chaling sites [53].

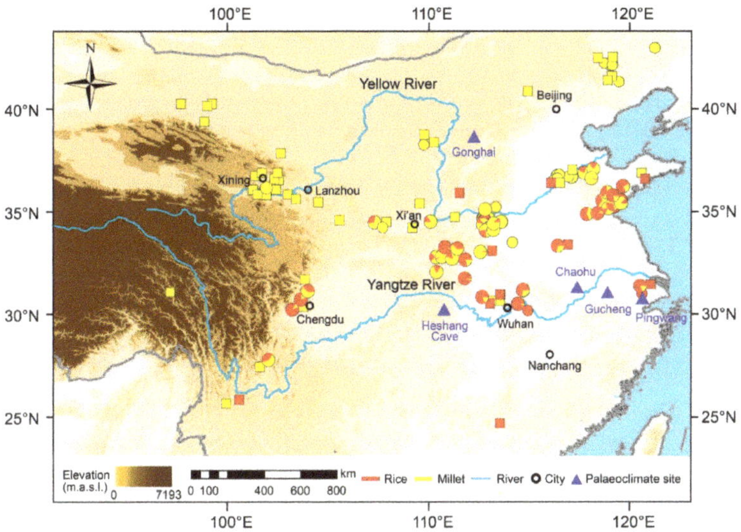

Figure 3. The spatial pattern of crop macro-fossils from sites dated between 5000–4000 BP in China. Squares represent sites without detailed archaeobotanical data; circles represent sites with detailed archaeobotanical data.

6. Spatial–Temporal Variation of Agriculture Patterns in Response to Climate Change in Neolithic China

The archaeobotanical evidence reveals a series of broad changes in the spatial patterns of agricultural development in Neolithic China. The overall area of rice cultivation appears to be larger than that of millet during 9000–7000 BP, and the boundary between these two traditional agricultural systems lay roughly east–west at ~34 °N (Figure 1). This boundary moved slightly southward to ~33 °N during 7000–5000 BP, as illustrated by the distribution of sites that yielded charred millet grains (Figure 2). During 5000–4000 BP, the boundary between rice and millet shifted to an approximate northeast–southwest direction. In east China, the northern limit for rice-based agriculture in the Shandong Peninsula moved to ~36 °N. In the Chengdu Plain of southwest China, the local cropping structure changed from a combination of broomcorn and foxtail millet to a combination of rice and foxtail millet at around 4700 BP [45]. Millet agriculture appears most dominant in the upper and middle valley of the Yellow River, the Hexi Corridor and the Yanshan-Liaoning area of northeast China (Figure 3).

Given the fact that the environmental conditions for the growth of the crops vary between species, broad changes in agriculture activities can be correlated with paleoclimate records. In order to better understand how the spatial–temporal variation of rice and millet developed, we compared the broad changes in agriculture activities with a number of key paleoclimate records in northern and central China. The record, with a ~20 year resolution precipitation reconstruction from Gonghai Lake (Figure 4f, [54]) in northern China, is indicative of gradually increasing precipitation from 14,600 to 7800 BP. Precipitation reached a maximum between 7800 and 5300 BP and decreased after 5300 BP. The generally high rainfall/moisture stage from ~8000 to ~5000 BP has also been recorded in other lake sediments and loess sections such as the Daihai Lake [55] and the Dadiwan section [56]. It is also possible to cross-check the change in precipitation with different kinds of paleoclimate records. For example, a large synthesis of precipitation based on 310 dates from 77 sites on the Loess Plateau shows that the paleosol probability density was relatively high from 8000 to 5000 BP, reflecting the relatively high moisture conditions during this period [57].

Precipitation and moisture records from the middle and lower reaches of the Yangtze river during the Holocene are relatively rare and more difficult to interpret. The pollen-based precipitation record from the Chaohu Lake [58], the Gucheng Lake [59] and the Pingwang Lake [60] in the lower Yangtze region suggest that precipitation reached its maximum between 10,000 and 7000 BP, after which it followed a broad decline with strong oscillations up to the present day (Figure 4d, [61]). Moreover, the stalagmite ARM/SIRM record from the Heshang Cave (Figure 4e, [62]) and the mass accumulation rates of hopanoids from the Dajiuhu Peat bog [62], often used to reflect the Holocene paleo-humidity variations of the middle reaches of the Yangtze River, also demonstrate that the climate became more humid between 11,000–7000 BP and 3000–1000 BP but was more arid and highly variable between 7000 and 3000 BP. Although a number of small inconsistencies concerning precipitation or moisture can be found in these records, the consensus is that there was relatively high precipitation and moisture from 10,000 to 7000 BP which decreased between 7000 and 5000 BP. Precipitation and moisture appear to have subsequently increased from 5000–4000 BP in comparison.

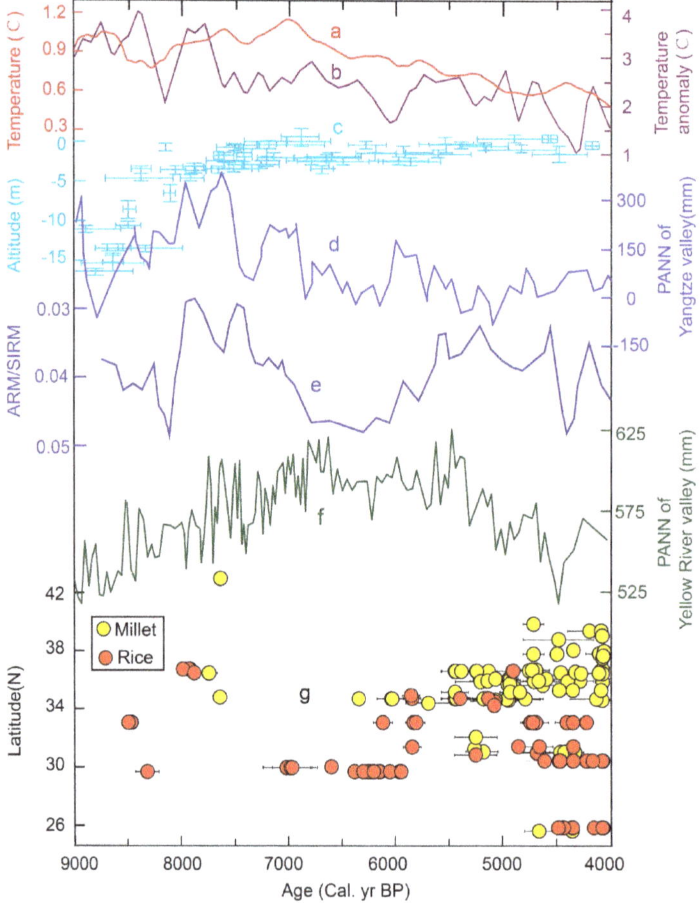

Figure 4. Comparison of (**a**,**b**) temperature anomaly in the Northern Hemisphere (30–90 °N) [15] and High Arctic [63]; (**c**) reconstructed sea-level change [64]; (**d**) pollen-based annual precipitation (PANN) in the Yangtze River valley [61]; (**e**) stalagmite ARM/SIRM record from the Heshang Cave [62]; (**f**) reconstructed PANN from Gonghai lake [54]; (**g**) latitudinal distribution of Neolithic sites with unearthed rice or millet sites between 9000–4000 BP.

We have considered the temperature anomaly in the Northern Hemisphere (30–90 °N) [15] and High Arctic [63] and pollen-based pollen-based annual precipitations (PANNs) for the Yangtze River region [61] and the Gonghai Lake [54], together with the stalagmite ARM/SIRM record from the Heshang Cave [62], as paleoclimate data, and have reconstructed the sea-level change [64] for comparison with the archaeobotanical results (Figure 4). Although the local paleoclimates of the region are complicated, the trend of the multiple paleoclimate records is largely valid on the regional scale. It was relatively wet in the Yangtze River valley and dry in the Yellow River valley during 9000–7000 BP and 5000–4000 BP, but these conditions changed to become completely opposite during 7000–5000 BP (Figure 4).

We suggest that the change to a relatively high temperature and precipitation during 9000–7000 BP (Figure 4a,b) provided ideal conditions for rice cultivation along the Yangtze River. Its growth requires both an appropriate temperature and adequate moisture. In comparison, millet (broomcorn and foxtail) are drought-tolerant and frost-sensitive and thus better adapted to the climate in northern China. However, the sea level transgression between 9000–7000 BP in the coastal plain of eastern China exerted considerable impact on the broad-spectrum economy based on fishing, hunting/gathering and agriculture along the lower reaches of the Yangtze River [65,66]. In the pre-Yangshao period, the expansion of millet seems to have been delayed in the Yellow River valley. The reason for this is still unclear but may be related to the low survival pressure mitigated by hunting/gathering. In this period, primitive agriculture was still in its infant stage, and hunting/gathering was the primary source of food supply in Yangtze River and Yellow River valleys [37,39].

During 7000–5000 BP, both precipitation and temperature began to decrease and the climate of the Yangtze River valley became dry and cool [15,61]. The sea level also began to lower at the same time (Figure 4c, [64,67]). The dry conditions and regression of the sea level after 7000 BP were disadvantageous for both fishing and hunting/gathering but provided an open landscape favorable to the rapid development of agriculture. During this time, the genetic characteristics of the rice remains from sites of the Hemudu culture (7000–5300 BP) tend to be stable and agricultural tools and pottery technology had also been significantly improved [68,69], both of which could also have facilitated rice cultivation and the expansion of this practice. On the other hand, powerful local societies such as Chengtoushan (c.6000 BP) and Liangzhu (5200–4300 BP) started to emerge in the middle and lower reaches of Yangtze River valley [70,71], indicating that the growth and aggregation of regional populations may have become increasingly dependent on agriculture [21]. Despite the drier and cooler climates, the changes in the sea level and human societies as a whole possibly triggered the major transition of food strategies to rice cultivation in the Yangtze River valley during 7000–5000 BP [21,66,72].

In the Yellow River valley, however, precipitation was evidently higher during 7000–5000 BP than 9000–7000 BP (Figure 4f, [54]). Thanks to the abundant water supply, millet agriculture became more intensified between 7000–6000 BP in the middle Yellow River valley [21,39]. This was also possibly related to the increasing size of the local population in northern China, which in return required more food supply and intensification in agriculture, particularly during the Yangshao period [73–75]. These human activities resulted in a sharp decline in forestation, making the area increasingly disadvantageous for hunting/gathering but favorable for the rise of millet cultivation [76]. The favorable climate might also facilitate rice cultivation in the Yellow River valley. Genetic evidence also shows a continuous movement of people from southern China to the Yellow River valey since the Yangshao period [42,77]. Meanwhile, the adoption of millet cultivation and its southward expansion, boosted by this climate change, is exemplified in the Chengdu Plain in upper Yangtze River valley between 6000–4700 BP, where millet became the major food supply [45].

During 5000–4000 BP, precipitation declined in the Yellow River valley but increased in the Yangtze River valley [54,61,62], while temperature followed a constant decline compared to 7000–5000 BP [15]. Climate deterioration in northern China appears to be one of the key reasons for the collapse of the Yangshao culture, which mainly relied on the supply of millet [54]. The relatively wet climate in the

Yangtze River valley during 5000–4000 BP [61] provided favorable conditions for rice cultivation, which might have promoted the transition of cropping patterns in the Chengdu Plain from millet agriculture to mixed rice–millet agriculture [45]. Meanwhile, mixed agriculture also began to thrive in the Huai River valley, located between the Yangtze River and the Yellow River [78]. The increasingly diversified farming activities considerably improved human adaptability to the widespread climate changes occurring between 5000 and 4000 BP [54,79].

In addition to millet and rice, during the second half of the Longshan period (ca. 4500–4000 BP), another important crop in Chinese prehistory—wheat—was introduced into the lower Yellow River valley [80]. As an exotic crop, it was initially not adopted as a major staple in Neolithic China. The introduction and cultivation of cold-tolerant barley and wheat greatly altered the cropping structures of northern China during the Bronze Age, especially in northwest China, where the altitude is much higher than in east China [44,81]. Agricultural innovation in the Tibetan Plateau featured the cultivation of barley and herding of sheep and yak, enabling local people to move into higher-elevation areas and settle in them permanently after 3600 BP when the climate changed to become cold and dry [44,82].

7. Conclusions

Archaeobotanical studies present the long and complicated trajectory of indigenous agricultural development in China. It is certainly not as simple (i.e., northern millet and southern rice) as noted in the historical documents but rather a dynamic process involving and responding to the key elements of climate change. The period 9000–4000 BP was characterized by the combination of rice-based agriculture in the Yangtze River valley and millet-based agriculture in the Yellow River valley, together with a series of variations in regional cropping patterns during different phases of the Neolithic Age. After 7000 BP, there was an important decline in temperature, which might have triggered the transformation of landscape and vegetation and promoted the transition from hunting/gathering to farming activities. The spatial–temporal variation of precipitation played an influential role in the shifting spatial patterns of farming activities during 6000–4000 BP, shedding more light on the issue of humans adapting to climate change in China during the Neolithic period prior to the adoption of exotic crops such as wheat and barley from the west.

Author Contributions: Data curation, R.L.; Investigation, R.L., L.Y. and F.L.; Methodology, F.L.; Supervision, G.D.; Visualization, L.Y.; Writing—original draft, R.L. and G.D.; Writing—review & editing, F.L. and R.L. All authors have read and agreed to the published version of the manuscript.

Funding: This research was funded by the National Natural Science Foundation of China (Grant Nos. 41671077 and 41825001) and supported by the 111 Project.

Acknowledgments: We would like to thank the three anonymous reviewers for their useful comments and Philly Howarth from the University of Oxford for editing the manuscript.

Conflicts of Interest: The authors declare no conflicts of interest.

References

1. Weiss, H.; Courty, M.-A.; Wetterstrom, W.; Guichard, F.; Senior, L.; Meadow, R.; Curnow, A. The Genesis and Collapse of Third Millennium North Mesopotamian Civilization. *Science* **1993**, *261*, 995–1004. [CrossRef]
2. Küper, R. Climate-Controlled Holocene Occupation in the Sahara: Motor of Africa's Evolution. *Science* **2006**, *313*, 803–807. [CrossRef]
3. Dong, G.; Liu, F.; Chen, F. Environmental and technological effects on ancient social evolution at different spatial scales. *Sci. China Earth Sci.* **2017**, *60*, 2067–2077. [CrossRef]
4. Wu, W.; Zheng, H.; Hou, M.; Ge, Q. The 5.5 cal ka BP climate event, population growth, circumscription, and the emergence of the earliest complex societies in China. *Sci. China Earth Sci.* **2017**, *61*, 134–148. [CrossRef]
5. Lespez, L.; Glais, A.; Lopez-Saez, J.-A.; le Drezen, Y.; Tsirtsoni, Z.; Davidson, R.; Biree, L.; Malamidou, D. Middle Holocene rapid environmental changes and human adaptation in Greece. *Quat. Res.* **2016**, *85*, 227–244. [CrossRef]

6. Wang, C.; Lu, H.; Gu, W.; Wu, N.; Zhang, J.; Zuo, X.; Li, F.; Wang, D.; Dong, Y.; Wang, S.; et al. The spatial pattern of farming and factors influencing it during the Peiligang culture period in the middle Yellow River valley, China. *Sci. Bull.* **2017**, *62*, 1565–1568. [CrossRef]

7. Cheng, Z.; Weng, C.; Steinke, S.; Mohtadi, M. Anthropogenic modification of vegetated landscapes in southern China from 6000 years ago. *Nat. Geosci.* **2018**, *11*, 939–943. [CrossRef]

8. Dong, G. A new story for wheat into China. *Nat. Plants* **2018**, *4*, 243–244. [CrossRef]

9. Diamond, J.; Bellwood, P. Farmers and Their Languages: The First Expansions. *Science* **2003**, *300*, 597–603. [CrossRef]

10. Gignoux, C.; Henn, B.M.; Mountain, J.L. Rapid, global demographic expansions after the origins of agriculture. *Proc. Natl. Acad. Sci. USA* **2011**, *108*, 6044–6049. [CrossRef]

11. Dong, G.; Yang, Y.; Han, J.; Wang, H.; Chen, F. Exploring the history of cultural exchange in prehistoric Eurasia from the perspectives of crop diffusion and consumption. *Sci. China Earth Sci.* **2017**, *60*, 1110–1123. [CrossRef]

12. Liu, X.; Jones, P.J.; Motuzaite-Matuzeviciute, G.; Hunt, H.V.; Lister, D.L.; An, T.; Przelomska, N.; Kneale, C.J.; Zhao, Z.; Jones, M.K. From ecological opportunism to multi-cropping: Mapping food globalisation in prehistory. *Quat. Sci. Rev.* **2019**, *206*, 21–28. [CrossRef]

13. Jia, X.; Dong, G.; Li, H.; Brunson, K.; Chen, F.; Ma, M.; Wang, H.; An, C.; Zhang, K. The development of agriculture and its impact on cultural expansion during the late Neolithic in the Western Loess Plateau, China. *Holocene* **2012**, *23*, 85–92. [CrossRef]

14. Yang, X.; Wu, W.; Perry, L.; Ma, Z.; Bar-Yosef, O.; Cohen, D.J.; Zheng, H.; Ge, Q. Critical role of climate change in plant selection and millet domestication in North China. *Sci. Rep.* **2018**, *8*, 7855. [CrossRef] [PubMed]

15. Marcott, S.A.; Shakun, J.D.; Clark, P.U.; Mix, A.C. A Reconstruction of Regional and Global Temperature for the Past 11,300 Years. *Science* **2013**, *339*, 1198–1201. [CrossRef]

16. Chen, F.; Yu, Z.; Yang, M.; Ito, E.; Wang, S.; Madsen, D.B.; Huang, X.; Zhao, Y.; Sato, T.; Birks, H.J.B.; et al. Holocene moisture evolution in arid central Asia and its out-of-phase relationship with Asian monsoon history. *Quat. Sci. Rev.* **2008**, *27*, 351–364. [CrossRef]

17. Chen, F.; Chen, J.; Holmes, J.; Boomer, I.; Austin, P.; Gates, J.B.; Wang, N.-L.; Brooks, S.J.; Zhang, J.-W. Moisture changes over the last millennium in arid central Asia: A review, synthesis and comparison with monsoon region. *Quat. Sci. Rev.* **2010**, *29*, 1055–1068. [CrossRef]

18. Lu, H.; Zhang, J.; Liu, K.-B.; Wu, N.; Li, Y.; Zhou, K.; Ye, M.; Zhang, T.; Zhang, H.; Yang, X.; et al. Earliest domestication of common millet (Panicum miliaceum) in East Asia extended to 10,000 years ago. *Proc. Natl. Acad. Sci. USA* **2009**, *106*, 7367–7372. [CrossRef]

19. Zhao, Z. New Archaeobotanic Data for the Study of the Origins of Agriculture in China. *Curr. Anthr.* **2011**, *52*, S295–S306. [CrossRef]

20. Zuo, X.; Lu, H.; Jiang, L.; Zhang, J.; Yang, X.; Huan, X.; He, K.; Wang, C.; Wu, N. Dating rice remains through phytolith carbon-14 study reveals domestication at the beginning of the Holocene. *Proc. Natl. Acad. Sci. USA* **2017**, *114*, 6486–6491. [CrossRef]

21. Zhao, Z.J. The formation of ancient Chinese agriculture—Evidence of plant remains unearthed from flotation. *Quat. Sci.* **2014**, *34*, 73–84.

22. Zhang, H.; Griffiths, M.L.; Chiang, J.C.H.; Kong, W.; Wu, S.-T.; Atwood, A.R.; Huang, J.; Cheng, H.; Ning, Y.; Xie, S. East Asian hydroclimate modulated by the position of the westerlies during Termination I. *Science* **2018**, *362*, 580–583. [CrossRef] [PubMed]

23. Zhang, C.; Pollard, A.M.; Rawson, J.; Huan, L.; Liu, R.; Tang, X. China's major Late Neolithic centres and the rise of Erlitou. *Antiquiti* **2019**, *93*, 588–603. [CrossRef]

24. Yang, Y.; Ren, L.; Dong, G.; Cui, Y.; Liu, R.; Chen, G.; Wang, H.; Wilkin, S.; Chen, F. Economic Change in the Prehistoric Hexi Corridor (4800–2200bp), North-West China. *Archaeometry* **2019**, *61*, 957–976. [CrossRef]

25. Liu, L.; Chen, X.C. *The Archaeology of China: From the Late Paleolithic to the Early Bronze Age*; SDX Joint Publishing Company Press: Beijing, China, 2012; ISBN 978-7-108-05901-7.

26. Yang, X.; Wan, Z.; Perry, L.; Lu, H.; Wang, Q.; Zhao, C.; Li, J.; Xie, F.; Yu, J.; Cui, T.; et al. Early millet use in northern China. *Proc. Natl. Acad. Sci. USA* **2012**, *109*, 3726–3730. [CrossRef]

27. Bestel, S.; Bao, Y.; Zhong, H.; Chen, X.; Liu, L. Wild plant use and multi-cropping at the early Neolithic Zhuzhai site in the middle Yellow River region, China. *Holocene* **2017**, *28*, 195–207. [CrossRef]

28. Wang, C.; Lu, H.; Gu, W.; Zuo, X.; Zhang, J.; Liu, Y.; Bao, Y.; Hu, Y. Temporal changes of mixed millet and rice agriculture in Neolithic-Bronze Age Central Plain, China: Archaeobotanical evidence from the Zhuzhai site. *Holocene* **2017**, *28*, 738–754. [CrossRef]

29. Zhao, Z.J. The study of the origin millet-new archaeobotany data and ecological analysis. *J. Chifeng Univ.* **2008**, *1*, 35–38.

30. Crawford, G.W.; Chen, X.X.; Luan, F.S. A Preliminary analysis on plant remains of the yuezhuang site in changqing district, Jinan City, Shandong Province. *Jianghan Archaeol.* **2013**, *2*, 107–116.

31. Zhao, Z.J.; Zhang, J.Z. Report on the analysis of the 2001 flotation of the Jiahu site. *Archaeology* **2009**, *8*, 84–93.

32. Zhang, J.Z.; Cheng, Z.J.; Lan, W.L.; Yang, Y.Z.; Luo, W.H.; Yao, L.; Yin, C.L. New progress in archaeobotany of Jiahu site, Wuyang, Henan province. *Archaeology* **2018**, *607*, 100–110.

33. Liu, C. Report on the analysis of the plant remains from Dadiwan. In *Site of Dadiwan in Qin'an*; Cultural Relics Press: Beijing, China, 2006; pp. 914–916.

34. Liu, X.; Hunt, H.V.; Jones, M.K. River valleys and foothills: Changing archaeological perceptions of North China's earliest farms. *Antiquity* **2009**, *83*, 82–95. [CrossRef]

35. Hu, Y.; Wang, S.; Luan, F.; Wang, C.; Richards, M.P. Stable isotope analysis of humans from Xiaojingshan site: Implications for understanding the origin of millet agriculture in China. *J. Archaeol. Sci.* **2008**, *35*, 2960–2965. [CrossRef]

36. Atahan, P.; Dodson, J.; Li, X.; Zhou, X.; Hu, S.; Chen, L.; Bertuch, F.; Grice, K. Early Neolithic diets at Baijia, Wei River valley, China: Stable carbon and nitrogen isotope analysis of human and faunal remains. *J. Archaeol. Sci.* **2011**, *38*, 2811–2817. [CrossRef]

37. Barton, L.; Newsome, S.D.; Chen, F.-H.; Wang, H.; Guilderson, T.P.; Bettinger, R.L. Agricultural origins and the isotopic identity of domestication in northern China. *Proc. Natl. Acad. Sci. USA* **2009**, *106*, 5523–5528. [CrossRef] [PubMed]

38. Liu, X.; Jones, M.K.; Zhao, Z.; Liu, G.; O'Connell, T.C. The earliest evidence of millet as a staple crop: New light on neolithic foodways in North China. *Am. J. Phys. Anthr.* **2012**, *149*, 283–290. [CrossRef] [PubMed]

39. Dong, G.; Zhang, S.; Yang, Y.; Chen, J.; Chen, F. Agricultural intensification, and its impact on environment during Neolithic Age in northern China. *Chin. Sci. Bull.* **2016**, *61*, 2913–2925. [CrossRef]

40. Zhejiang institute of cultural relics and archaeology. *A Report on Archaeological Excavation of a Neolithic Site, Hemudu*; Cultural Relics Press: Beijing, China, 2003; ISBN 9787501014521.

41. Fuller, D.Q.; Qin, L.; Zheng, Y.; Zhao, Z.; Chen, X.; Hosoya, L.A.; Sun, G.-P. The Domestication Process and Domestication Rate in Rice: Spikelet Bases from the Lower Yangtze. *Science* **2009**, *323*, 1607–1610. [CrossRef]

42. Li, X.; Zhou, X.; Zhou, J.; Dodson, J.; Zhang, H.; Shang, X. The earliest archaeobiological evidence of the broadening agriculture in China recorded at Xishanping site in Gansu Province. *Sci. China Ser. D Earth Sci.* **2007**, *50*, 1707–1714. [CrossRef]

43. Sergusheva, E.A.; Vostretsov, Y.E. The Advance of Agriculture in the Coastal Zone of East Asia. In *From Foragers to Farmers*; Fairbairn, A., Weiss, E., Eds.; Oxbow Books: Oxford, UK, 2009; pp. 205–219, ISBN 9781842173541.

44. Dong, G.; Zhang, D.; Liu, X.; Liu, F.; Chen, F.; Jones, M. Response to Comment on Agriculture facilitated permanent human occupation of the Tibetan Plateau after 3600 B.P. *Science* **2015**, *348*, 872. [CrossRef]

45. D'Alpoim, G.J.M. Rice, social complexity, and the spread of agriculture to the Chengdu Plain and Southwest China. *Rice* **2011**, *4*, 104–113. [CrossRef]

46. Nasu, H.; Momohara, A.; Yasuda, Y.; He, J. The occurrence and identification of *Setaria italica* (L.) P. Beauv. (foxtail millet) grains from the Chengtoushan site (ca. 5800 cal B.P.) in central China, with reference to the domestication centre in Asia. *Veg. Hist. Archaeobotany* **2006**, *16*, 481–494. [CrossRef]

47. Zhou, X.; Xiaoqiang, L.; Dodson, J.; Keliang, Z. Rapid agricultural transformation in the prehistoric Hexi corridor, China. *Quat. Int.* **2016**, *426*, 33–41. [CrossRef]

48. Li, S.C. *Prehistoric Culture Evolution in Northwest China*; Cultural Relics Press: Beijing, China, 2009; ISBN 9787501026555.

49. D'Alpoim-Guedes, J.; Lu, H.; Li, Y.; Spengler, R.N.; Wu, X.; Aldenderfer, M.S. Moving agriculture onto the Tibetan plateau: The archaeobotanical evidence. *Archaeol. Anthr. Sci.* **2013**, *6*, 255–269. [CrossRef]

50. D'Alpoim, G.J. Rethinking the spread of agriculture to the Tibetan Plateau. *Holocene* **2015**, *25*, 1498–1510. [CrossRef]

51. Li, H.; Zuo, X.; Kang, L.; Ren, L.; Liu, F.; Liu, H.; Zhang, N.; Min, R.; Liu, X.; Dong, G. Prehistoric agriculture development in the Yunnan-Guizhou Plateau, southwest China: Archaeobotanical evidence. *Sci. China Earth Sci.* **2016**, *59*, 1562–1573. [CrossRef]

52. Martello, R.D.; Min, R.; Stevens, C.J.; Higham, C.; Higham, T.; Qin, L.; Fuller, D.Q. Early agriculture at the crossroads of China and Southeast Asia: Archaeobotanical evidence and radiocarbon dates from Baiyangcun, Yunnan. *J. Archaeol. Sci. Rep.* **2018**, *20*, 711–721. [CrossRef]

53. Yang, X.; Chen, Q.; Ma, Y.; Li, Z.; Hung, H.-C.; Zhang, Q.; Jin, Z.; Liu, S.; Zhou, Z.; Fu, X. New radiocarbon and archaeobotanical evidence reveal the timing and route of southward dispersal of rice farming in south China. *Sci. Bull.* **2018**, *63*, 1495–1501. [CrossRef]

54. Chen, F.; Xu, Q.; Chen, J.; Birks, H.J.B.; Liu, J.; Zhang, S.; Jin, L.; Kandasamy, S.; Telford, R.J.; Cao, X.; et al. East Asian summer monsoon precipitation variability since the last deglaciation. *Sci. Rep.* **2015**, *5*, 11186. [CrossRef]

55. Xiao, J.; Xu, Q.; Nakamura, T.; Yang, X.; Liang, W.; Inouchi, Y. Holocene vegetation variation in the Daihai Lake region of north-central China: A direct indication of the Asian monsoon climatic history. *Quat. Sci. Rev.* **2004**, *23*, 1669–1679. [CrossRef]

56. Feng, Z.; Tang, L.; Wang, H.; Ma, Y.; Liu, K.-B. Holocene vegetation variations and the associated environmental changes in the western part of the Chinese Loess Plateau. *Palaeogeogr. Palaeoclim. Palaeoecol.* **2006**, *241*, 440–456. [CrossRef]

57. Wang, H.; Chen, J.; Zhang, X.; Chen, F. Palaeosol development in the Chinese Loess Plateau as an indicator of the strength of the East Asian summer monsoon: Evidence for a mid-Holocene maximum. *Quat. Int.* **2014**, *334*, 155–164. [CrossRef]

58. Chen, W.; Wang, W.-M.; Dai, X.-R. Holocene vegetation history with implications of human impact in the Lake Chaohu area, Anhui Province, East China. *Veg. Hist. Archaeobotany* **2008**, *18*, 137–146. [CrossRef]

59. Yang, X.D.; Wang, S.M.; Tong, G.B. Character of palynology and changes of monsoon climate over the last 10,000 years in Gucheng Lake, Jiangsu Province. *Acta. Bot. Sin.* **1996**, *38*, 576–581.

60. Innes, J.B.; Zong, Y.Q.; Wang, Z.H.; Chen, Z.Y. Climatic and palaeoeco-logical changes during the mid- to late Holocene transition in eastern china: High-resolution pollen and non-pollen palyno-morph analysis at Pingwang, Yangtze coastal lowlands. *Quat. Sci. Rev.* **2014**, *99*, 164–175. [CrossRef]

61. Li, J.; Dodson, J.; Yan, H.; Wang, W.; Innes, J.B.; Zong, Y.; Zhang, X.; Xu, Q.; Ni, J.; Lu, F. Quantitative Holocene climatic reconstructions for the lower Yangtze region of China. *Clim. Dyn.* **2017**, *50*, 1101–1113. [CrossRef]

62. Xie, S.; Evershed, R.P.; Huang, X.; Zhu, Z.; Pancost, R.D.; Meyers, P.; Gong, L.; Hu, C.; Huang, J.; Zhang, S.; et al. Concordant monsoon-driven postglacial hydrological changes in peat and stalagmite records and their impacts on prehistoric cultures in central China. *Geology* **2013**, *41*, 827–830. [CrossRef]

63. Lecavalier, B.S.; Fisher, D.A.; Milne, G.A.; Vinther, B.; Tarasov, L.; Huybrechts, P.; Lacelle, D.; Main, B.; Zheng, J.; Bourgeois, J.; et al. High Arctic Holocene temperature record from the Agassiz ice cap and Greenland ice sheet evolution. *Proc. Natl. Acad. Sci. USA* **2017**, *114*, 5952–5957. [CrossRef]

64. Sun, Y. Holocene Sea Level Change in the Western North Pacific Marginal Seas and Coastal Responses to Recent Sea Level Change in the Deep Bay Wetlands. Ph.D. Thesis, The University of Hong Kong Libraries, Hongkong, China, 2017.

65. Lai, Y.; Zhang, J.Z.; Yin, R.C. On Instruments of Production and Economic Structure of Jiahu Site in Wuyang. *Cul. Relics Cent. China* **2009**, 22–28.

66. Zheng, H.; Zhou, Y.; Yang, Q.; Hu, Z.; Ling, G.; Zhang, J.; Gu, C.; Wang, Y.; Cao, Y.; Huang, X.; et al. Spatial and temporal distribution of Neolithic sites in coastal China: Sea level changes, geomorphic evolution and human adaption. *Sci. China Earth Sci.* **2017**, *61*, 123–133. [CrossRef]

67. Xiong, H.; Zong, Y.; Qian, P.; Huang, G.; Fu, S. Holocene sea-level history of the northern coast of South China Sea. *Quat. Sci. Rev.* **2018**, *194*, 12–26. [CrossRef]

68. Zhang, J.Z.; Chen, C.F.; Yang, Y.Z. Origins and Early Development of Agriculture in China. *J. Nat. Mus. China* **2014**, *126*, 6–16.

69. Peng, B. Agricultural remains of early rice cultivation in China and related problems. *Agricult. Archaeol.* **2016**, *143*, 50–55.

70. Hunan Institute of Cultural Relics and Archaeology. *Excavation of Neolithic Site in Chengtou Mountain, Lixian County*; Cultural Relics Press: Beijing, China, 2007; pp. 164–167, ISBN 9787501017515.

71. Zhejiang institute of cultural relics and archaeology. Excavation of Liangzhu ancient city site from 2006 to 2007 in Yuhang district, Hangzhou city. *Archaeology* **2008**, *7*, 1–8.

72. Patalano, R.; Wang, Z.; Leng, Q.; Liu, W.; Zheng, Y.; Sun, G.; Yang, H. Hydrological changes facilitated early rice farming in the lower Yangtze River Valley in China: A molecular isotope analysis. *Geology* **2015**, *43*, 639–642. [CrossRef]

73. Hosner, D.; Wagner, M.; Tarasov, P.E.; Chen, X.; Leipe, C. Spatiotemporal distribution patterns of archaeological sites in China during the Neolithic and Bronze Age: An overview. *Holocene* **2016**, *26*, 1576–1593. [CrossRef]

74. Dong, G.; Li, R.; Lu, M.; Zhang, D.; James, N. Evolution of human–environmental interactions in China from the Late Paleolithic to the Bronze Age. *Prog. Phys. Geogr. Earth Environ.* **2019**, *44*, 233–250. [CrossRef]

75. Wang, J.H. A Preliminary Study on the Population Size and Related Issues in the Yangshao Period of Shaanxi Province. *Archaeol. Cult. Relics* **2009**, 26–35.

76. Tang, L.Y.; An, C.B. The vegetation history and aridevent referred by pollen records on the Loess Plateau in Longzhong area. *Proc. Nat. Sci.* **2007**, *17*, 1371–1382.

77. Ning, C.; Li, T.; Wang, K.; Zhang, F.; Li, T.; Wu, X.; Gao, S.; Zhang, Q.; Zhang, H.; Hudson, M.J.; et al. Ancient genomes from northern China suggest links between subsistence changes and human migration. *Nat. Commun.* **2020**, *11*, 2700. [CrossRef]

78. Yang, Y.; Cheng, Z.; Li, W.; Yao, L.; Li, Z.; Luo, W.; Yuan, Z.; Zhang, J.; Zhang, J. The emergence, development, and regional differences of mixed farming of rice and millet in the upper and middle Huai River Valley, China. *Sci. China Earth Sci.* **2016**, *59*, 1779–1790. [CrossRef]

79. Myers, C.G.; Oster, J.L.; Sharp, W.D.; Bennartz, R.; Kelley, N.P.; Covey, A.K.; Breitenbach, S.F. Northeast Indian stalagmite records Pacific decadal climate change: Implications for moisture transport and drought in India. *Geophys. Res. Lett.* **2015**, *42*, 4124–4132. [CrossRef]

80. Long, T.; Leipe, C.; Jin, G.; Wagner, M.; Guo, R.; Schröder, O.; Tarasov, P.E. The early history of wheat in China from 14C dating and Bayesian chronological modelling. *Nat. Plants* **2018**, *4*, 272–279. [CrossRef] [PubMed]

81. D'Alpoim-Guedes, J.; Lu, H.; Hein, A.; Schmidt, A.H. Early evidence for the use of wheat and barley as staple crops on the margins of the Tibetan Plateau. *Proc. Natl. Acad. Sci. USA* **2015**, *112*, 5625–5630. [CrossRef]

82. Dong, G.; Ren, L.; Jia, X.; Liu, X.; Dong, S.; Li, H.; Wang, Z.; Xiao, Y.; Chen, F. Chronology, and subsistence strategy of Nuomuhong Culture in the Tibetan Plateau. *Quat. Int.* **2016**, *426*, 42–49. [CrossRef]

Article

Isotopic Results Reveal Possible Links between Diet and Social Status in Late Shang Dynasty (ca. 1250–1046 BC) Tombs at Xiaohucun, China

Ning Wang [1,*], Lianmin Jia [2], Yi Si [3] and Xin Jia [4]

[1] School of History, Culture and Tourism, Jiangsu Normal University, Xuzhou 221116, China
[2] Henan Provincial Institute of Cultural Heritage and Archaeology, Zhengzhou 450000, China; jialianminhn@gmail.com
[3] School of History and Culture, Henan University, Kaifeng 475001, Henan, China; 10020113@vip.henu.edu.cn
[4] School of Geography, Nanjing Normal University, Nanjing 210023, China; jiaxin@njnu.edu.cn
* Correspondence: wanging3@gmail.com; Tel.: +86-0516-83536381

Received: 4 March 2020; Accepted: 23 April 2020; Published: 29 April 2020

Abstract: Here, we present evidence of possible links between diet and social status using carbon and nitrogen stable isotope ratios at the site of Xiaohucun in the Central Plains, China. This pilot study from a rescue excavation yielded humans (n = 12) identified to the late Shang Dynasty (ca. 1250–1046 BC), which was a warm climatic period. The population consumed a predominately C_4 diet (millets) and no difference was observed between the $\delta^{13}C$ results of individuals (n = 7) buried with (−9.1 ± 2.8‰) and without (n = 5) bronze vessels (−8.2 ± 0.7‰). However, individuals buried with bronze vessels (10.3±1.6‰) were found to have significantly higher $\delta^{15}N$ values (one-way ANOVA; $p = 0.015$) compared to individuals buried without bronze vessels (8.0 ± 0.9‰), providing evidence that possible elite members consumed more animal protein (dog, pig, cow, sheep/goat). Isotopic results were also examined for social status in relation to the number of burial coffins that an individual had: double (n = 6), single (n = 3), or no coffin (n = 3). No difference was found in the $\delta^{13}C$ values, but variations were observed in the $\delta^{15}N$ values: double coffin (10.2 ± 1.7‰) > single coffin (8.8 ± 1.8‰) > no coffin (8.0 ± 1.3‰), again possibly showing increased animal protein consumption linked to social status. Finally, isotopic results and status were studied by looking at the number of coffins and tomb size. Again, no correlation was observed for the $\delta^{13}C$ results, but a strong linear correlation ($R^2 = 0.85$) was observed for the $\delta^{15}N$ values of the individuals buried in two coffins vs. tomb size.

Keywords: human diet; hierarchy; bronze age; carbon and nitrogen stable isotope ratios

---

1. Introduction

Ancient China was a complex and highly socially stratified society [1–4]. Vast disparities existed between the nobility and the common people in areas such as rights, ownerships, diet, customs, behavior etc. These differences were chronicled in various historical works such as: Li Ji "礼记" which described how criminal law did not apply to senior officials and that etiquette did not apply to common people [5]. This social hierarchy of the living was also extended to the treatment of the dead, and the phrase: "Honor the dead as the living", recorded by Xun Zi "荀子", was and still is an important concept that is intertwined through the fabric of Chinese society [6,7]. Social status could also be maintained in the afterlife by the size and the scale of the tomb, and the quantity and quality of the grave goods interred with an individual [8–12].

Stable isotope ratio analysis has been successfully applied to examine dietary patterns in past populations from many societies across the globe [13–24]. Isotopic results of bone collagen primarily reflect the protein component of the diet averaged over the entire lifetime of an individual, including a large portion of collagen synthesized during later childhood and adolescence [25,26]. Briefly, stable isotope ratios are defined as the ratio of the heavier to the lighter isotope (e.g., $^{13}C/^{12}C$ or $^{15}N/^{14}N$) and are compared in terms of δ values in parts per 1000 or "per mil" (‰) in relation to internationally defined standards for carbon (Vienna Pee Dee Belemnite, VPDB) and nitrogen (ambient inhalable reservoir, AIR) [27]. In Chinese archaeological research, $\delta^{13}C$ measurements of human and animal collagen allow for an examination of the contribution of C_3 (rice, wheat, barley etc.) and C_4 foods (millets) to the diet. These studies have been vital for the reconstruction of a coarse time scale for the spread of different forms of agriculture in China [28–35] and for understanding animal husbandry practices [36–39]. The $\delta^{15}N$ results can be used as an estimation of the trophic level of a human or animal in a foodweb and are based on the observation of an increase of about 3‰–5‰ from the food to the consumer tissue [14,40,41]. Thus, levels of animal protein consumption can be examined with nitrogen isotope ratios. Modern human studies have found higher $\delta^{15}N$ values in omnivores compared to vegeterians and vegans [42,43]. However, an in-depth discussion on the intricacies of stable isotope ratios to reconstruct past diets is beyond the focus of this paper and the reader should consult the reviews of Katzenberg [44], Ambrose and Krigbaum [45], and Lee-Throp [46].

Increasingly, this technique is used to directly document dietary differences between social classes, e.g., elites vs. common people [47–52], and the reader is directed to consult Twiss [53] for a review of food and social diversity in archaeological and isotopic research. However, relatively little research has focused on the use of stable isotope ratios to directly determine dietary differences related to social status in China [54–58]. It is widely accepted that the Shang Dynasty (1600–1046 BC) is an early Chinese era based on abundant written records and the rich archaeological evidence [59–64]. Thus, this period is an extremely important phase in the development of the earliest Chinese civilization, and the formation of the ritual systems of power, class, and hierarchy [3]. Here, we present results of a small pilot study that examines dietary patterns related to social status for burials that date to the late Shang Dynasty (ca. 1250–1046 BC). Humans (n = 12) from the site of Xiaohucun and animals (n = 11) from the nearby contemporaneous site of Guandimiao in Henan Province were available for study due to a rescue excavation. The results of this research will be the focus of this work [65].

2. Background of Ritual Systems in Ancient China

China has a long and vibrant tradition of organized rituals and hierarchy, and this is especially evident for dining practices as well as the type of food consumed [66–70]. A separate system of dining was very popular for the elite members of society. At dinner, individuals kneeled next to an Aiji "案基" (a kind of little table) and a selection of tableware was placed next to each person for holding cereals, meat, water, beverages, and liquor. The type, quality, and number of the dishes served were determined by the status and the age of the diners. Elders and people of high social status had the right to use more bronze vessels [69,71]. According to written accounts and archaeological data, the most important tableware in China were Ding "鼎" (an ancient vessel for cooking or holding meat) and Gui "簋" (an ancient vessel for holding grains), and a combination of both is the Chinese characters for banquet "飨宴", which was a symbol associated with nobility [72] (Figure 1). The number of Ding and Gui that a person could use in life was consistent with their social status, and this was also reflected in the combination of these items that were buried with the dead (Table 1). In addition, the consumption of meat was a privilege of the nobility and elders of a family before the Qin Dynasty (Table 1). The work Li Ji "礼记" recorded that "If there is not an important reason, the princes shouldn't kill cattle to eat, senior officials shouldn't kill sheep to eat, junior officers shouldn't kill dogs and pigs to eat, and the common people should not eat meat [5]. Thus, meat was a luxury dietary item that was generally reserved for elites and only for special occasions such as festivals or banquets. In addition, analysis of ceramics from the early Shang Dynasty site of Yanshi determined different structures and behaviors of dining between the palace elites and the individuals that made pottery [73]. Reinhart determined that the Yanshi elites engaged in large-scale exclusionary feasting in contrast to the simpler and home cooked meals of the potters.

Figure 1. Bronze ritual vessels discovered in tomb M22 of Xiaohucun site (**a** = Ding, **b** = Gui, **c** = All bronze grave goods).

There were also strict rules for the maintenance of social status for the deceased in ancient China. Tomb size and burial depth (representing wealth and labor consumption) and the quantity and quality of the coffins used were some of the most important criteria reflecting social status (Table 1) [74,75]. The writings of Xun Zi "荀子" described that the number of coffins used for the Emperor was seven, for a prince was five, for senior nobility was three, and for junior nobility was two [6]. Furthermore, there were strict codes for the materials, size, thickness, as well as the internal and external decoration patterns of coffins, and this was related to hierarchy in Chinese society (Table 1). In addition, the type and quality of grave goods was an important manifestation of the status of the deceased. At least from the Zhou Dynasties, the combination of the number of bronze ritual vessels (Ding and Gui) played a very important role in the status of a tomb owner (Table 1) [3,72]. However, it is important to note that there is a chronological lag between these historical sources and the Shang period and that these works might be biased by the views of the writers. Thus, caution and some skepticism are advised in the use of these textual sources, and they should not be viewed as undisputed fact.

Table 1. Summary of selected differences based on social class in ancient China. This information is referenced from LiJi "礼记" [5] except for "coffin number" which was referenced from Xun Zi "荀子" [6]. Note: Readers are advised that care must be taken with the accuracy of this information as it is compiled from historical sources which could be subject to the biases of the writers.

Status	Diet		Sacrifice	Travel			Burial		
	Tableware and Courses per Meal	Meat in Diet	Bronze Vessels (Ding) in Ritual Activities	Carriage Number	Horse per Carriage	Coffin Number	Coffin Material	Ding + Gui Combination
Emperor	26	Yes	7	NA	6	7	N/A	9 & 8
Prince	12–16	Yes	5	7	4	5	Pine	7 & 6
Senior Nobility	8	Yes	4	5	3	3	Cypress	5 & 4
Junior Nobility	6	Yes	1	3	2	2	Miscellaneous	3 & 2
Common People	3–6	Only older people	N/A	N/A	1	N/A	N/A	N/A

3. Site of Xiaohucun, Henan Province, China

The Xiaohucun site, dating to the late Shang (ca. 1250–1046 BC) and Western Zhou Dynasty (ca. 1046–771 BC), was the focus of a partial rescue excavation in 2006 by the Henan Provincial Institute of Cultural Relics and Archaeology. The site is situated to the northeast of Xiaohu Village in Xingyang City, which is about 20 km from Zhengzhou, the capital of Henan Province, China (Figure 2) [76]. This unique location is situated in the core of the Central Plains, which was the origin of many early civilizations of China, and also the political and economic center of the early Shang Dynasty. Based on archaeobotanical and isotopic research, the inhabitants of the Central Plains mainly relied on millet cultivation (rice and wheat were also present to some extent) and animal husbandry (pig, cattle, etc.) for subsistence [1,3,34,66,77–80].

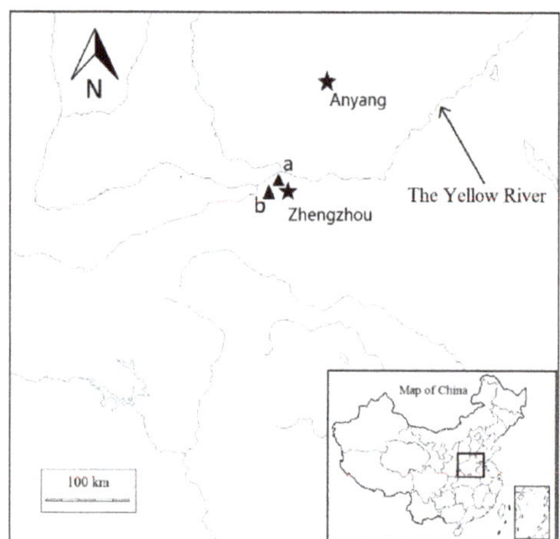

Figure 2. Map of China showing the location of the (**a**) Xiaohucun site and (**b**) Guandimiao site. Note: Zhengzhou is the capital of Henan Province.

A total of 58 tombs from the late Shang Dynasty (ca. 1250–1046 BC) were excavated from an area of 400 × 200 meters. All tombs are pits and rectangular in shape, and most had a platform consisting of: inner coffin, outer coffin, waste pit, sacrificial dogs, and various types of grave goods. Unfortunately, many tombs have been looted and destroyed and only 21 of them could be studied in any detail. According to the preliminary report from this site, all the 21 tombs can be divided into three styles, equivalent to the phases of the Yin Ruins II, III, and IV [81]. No radiocarbon dates are available from the site, but examination of the tomb style, grave goods, and the writing on the bronze vessels suggest that the cemetery was a family plot that belonged to the *She* "舌" family from the late Shang Dynasty. Many of the individuals were from the junior nobility class, but a number of common people were also interred here.

4. Materials and Methods

For the 21 tombs at Xiaohucun, bone samples consisting of long bone fragments were obtained from 12 late Shang Dynasty individuals for stable isotope ratio analysis. Osteological analysis, including sex determination and age estimation, was carried out according to standard methods [82]. Unfortunately, due to poor preservation and looting many of the skeletons were destroyed or incomplete and only 5 out of 12 individuals could be positively identified to gender and 9 out of 12 aged (Table 2). In addition, since Xiaohucun was a cemetery site, faunal samples were not available for study. However, we were able to obtain animal bones (n = 11) from the nearby (15 km) and contemporaneous site of Guandimiao (ca. 1250–771 BC) which serves as a baseline estimation for the human diets (Table 3).

Collagen was extracted at the Key Laboratory of Vertebrate Evolution and Human Origins of the Chinese Academy of Sciences, Institute of Vertebrate Paleontology and Paleoanthropology in Beijing, China, using the protocol outlined by Richards and Hedges [83]. The extracted collagen was well preserved and the majority of samples had collagen yields of over 1% and C: N between 3.0–3.2 (11/11 animals; 12/12 humans), which is indicative of collagen suitable for isotopic analysis [84]. The samples were measured with an Isoprime 100 IRMS coupled with Elementar Vario. Standard material for testing the carbon and nitrogen content was sulfonamides. For every 10 samples, we interpolated one IEAE-CH-6, IEAE-N-2, and IEAE-600 to make data corrections. The measurement precision for $\delta^{13}C$ and $\delta^{15}N$ results is ±0.2‰. SPSS 20.0 and Origin 8.0 were used for statistical analysis.

Table 2. Isotopic results and sample information for all humans from the Xiaohucun site, Henan Province, China.

Lab ID	Location	Sex	Age	Open Size (m²)	Depth (m)	Coffin Type	# Grave Goods	Bronze Vessels	# Bronze Vessels	# Jade & other Objects	δ^{13}C (‰)	δ^{15}N (‰)	%C	%N	C:N
1	M8	Male	45–50	7.35	4.30	double	6	yes	6	-	−8.8	12.8	29.9	11.5	3.0
2	M24	Male	55 ±	5.52	1.80	double	8	yes	7	1	−8.1	10.7	38.6	14.0	3.2
3	M27	?	55 ±	5.61	2.60	double	8	yes	3	5	−7.8	10.3	41.0	14.9	3.2
4	M30	?	?	4.57	2.00	double	7	yes	6	1	−15.2	7.6	42.1	15.3	3.2
5	M39	?	40-45	3.92	1.00	none	-	no	-	-	−8.6	6.7	40.2	14.9	3.1
6	M46	?	35 ±	2.60	0.90	single	-	no	-	-	−8.8	7.5	39.7	14.4	3.2
7	M47	?	25–30	1.92	1.60	single	-	no	-	-	−7.5	8.2	38.4	14.1	3.2
8	M52	?	?	6.80	3.58	double	11	yes	5	4 (1 cowrie shell & 1 wooden object)	−6.4	10.9	41.4	15.2	3.2
9	M89	?	50–55	2.30	0.70	none	1	no	-	1	−7.4	8.2	41.2	15.2	3.2
10	M90	Male	40–45	3.95	0.66	single	4	yes	2	1 (1 cowrie shell)	−9.3	10.8	39.6	14.7	3.2
11	M105	Male	?	4.91	2.25	double	5	yes	4	1	−8.0	9.2	40.0	14.8	3.1
12	M116	Female	45 ±	1.54	1.40	none	1	no	-	(1 cowrie shell)	−8.5	9.2	37.5	13.5	3.2

Table 3. Isotopic results and sample information for all fauna from the Guandimiao site, Henan Province, China.

Lab ID	Location	Species	δ^{13}C (‰)	δ^{15}N (‰)	%C	%N	C:N
a 1	H28	Pig (Sus scrofa domestica)	−8.2	7.8	35.1	13.0	3.2
a 2	H26	Pig (Sus scrofa domestica)	−9.9	8.6	31.6	12.0	3.1
a 3	H26	Deer (Cervus nippon)	−20.5	4.9	36.2	13.0	3.2
a 4	H741	Pig (Sus scrofa domestica)	−6.7	6.6	41.8	15.0	3.2
a 5	H932	Pig (Sus scrofa domestica)	−9.3	7.6	41.0	14.8	3.2
a 6	H932	Dog (Canis lupus familiaris)	−7.6	8.6	37.3	13.7	3.2
a 7	H1309	Cattle (Bos primigenius taurus)	−9.0	5.7	41.9	15.2	3.2
a 8	H1250	Pig (Sus scrofa domestica)	−11.6	8.8	22.2	9.0	2.9
a 9	H1251	Deer (Cervus nippon)	−15.0	7.6	40.0	14.3	3.3
a 10	J20	Sheep/Goat (Caprinae)	−10.7	7.5	39.8	14.3	3.2
a 11	G7	Dog (Canis lupus familiaris)	−7.5	7.6	37.4	13.8	3.2

5. Results

Sample information and the δ^{13}C and δ^{15}N values for humans and animals are presented in Tables 2 and 3. Given the work at the Xiaohucun site was a rescue excavation, and many tombs were looted, we fully acknowledge that the number of humans available for study was relatively small, but this collection serves as a pilot study and the first opportunity to gain a glimpse into possible isotopic dietary differences related to status during the late Shang Dynasty (ca. 1250–1046 BC) in the Central Plains of China.

5.1. Faunal Isotope Results

In Figure 3 the isotopic results of the humans and animals are presented. The δ^{13}C values for pigs (n = 5), dogs (n = 2), cow (n = 1), sheep/goat (n = 1) range from −6.7‰ to −11.6‰, which indicates that all of these animals had a diet predominately, if not exclusively, based on C_4 dietary protein sources. Past archaeological and isotopic research has revealed that millet agriculture was important to the residents of the Central Plains [1,28,77,78,85], and that the livestock were mostly consuming millet or its byproducts during the Shang Dynasty [39]. In contrast to the domestic animals, the two deer had δ^{13}C values that indicated either a mixed C_3/C_4 (−15.0‰) or exclusive C_3 (−20.5‰) diet. The finding of a deer with the mixed diet is interesting as this could suggest that this animal lived near the settlement and grazed on the millet fields or that it was possibly kept as pet. Similar results for deer were found at the Xia Dynasty (2070–1600 BC) site of Xinzhai, also in Henan Provence, and these ^{13}C-enriched deer were thought to have been possibly raised for the purposes of hunting by the elites of the society [86]. The dog (8.1‰), pig (7.9 ± 0.9‰), and sheep/goat (7.5‰) all had similar δ^{15}N values, suggesting feeding at a similar trophic level, whereas the deer (6.3‰) and cow (5.7‰) were lower. However, this site was closed to the Yellow river, so we could not eliminate the possibility of ingesting freshwater or marine fish resources by human.

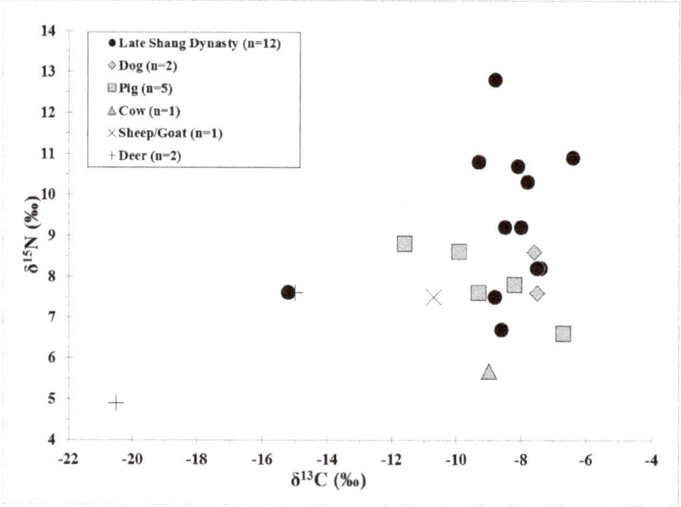

Figure 3. δ^{13}C and δ^{15}N results for humans from the Xiaohucun site and animals from the Guandimiao site Henan Province, China.

5.2. Human Isotope Results

The human δ^{13}C results range from −15.2‰ to −6.4‰ and show that most individuals had a predominately C$_4$ diet based on millet (Figure 3). The exception was individual M30 (−15.2‰) who had a mixed C$_3$/C$_4$ diet that was likely a combination of rice and/or wheat and millet [79,87]. The human δ^{15}N results range from 6.7‰ to 12.8‰, indicating there was likely significant individual variation in the consumption of protein at the site. A comparison of the human and faunal isotopes values indicate that pigs, dogs, cattle, and sheep/goats were all likely dietary protein sources, while deer played a minor role (Figure 3). Thus, the diet of the individuals was predominately millet based with some possible inputs of rice and/or wheat, but there were large differences in δ^{15}N values that we ascribe to variations in animal protein consumption, as discussed in the next section.

6. Discussion

6.1. Burials and the Classification of Social Status during the Late Shang Dynasty

In order to determine a possible correlation between social status and dietary differences at the Xiaohucun site, it is important to understand how social status may have influenced the tomb type and value. In ancient China, tombs and burials can be graded and valued based on the human labor costs and materials used for construction, the size and shape of the structure, as well as the number and type of graves goods reflecting symbols of power and wealth interred with individuals [55,71]. Since the shapes of all the late Shang Dynasty burials at Xiaohucun were rectangular, the preliminary grading scale of the 12 tombs was classified by number of coffins, tomb depth, and size, and by the presence of bronze funerary objects and this information is listed in Table 1 and plotted in Figure 4. Tomb sizes ranged from 1.54 m^2 to 7.35 m^2 (4.25 ± 1.9 m^2, n = 12), the burial depth was between 0.66 m to 4.3 m (1.93 ± 1.1 m, n = 12), which could suggest that the cemetery was under a type of burial management based on social hierarchy. More importantly, the burial depth and tomb size were found to have a linear correlation (R^2 = 0.67, *p* = 0.001) with clear differences between the number of coffins used and whether a burial contained bronze vessels.

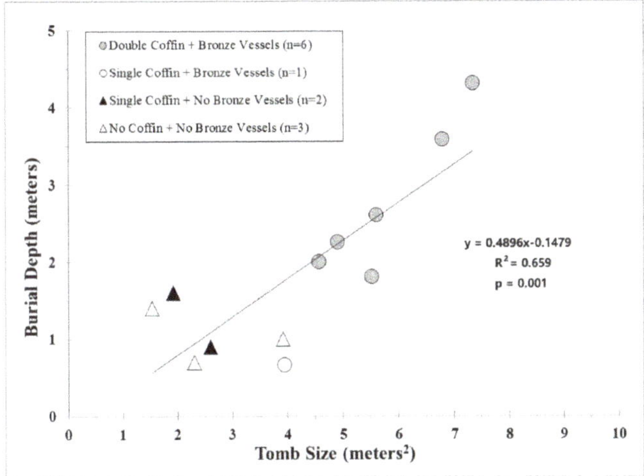

Figure 4. Plot of burial depth (m) vs. tomb size (m^2) for individuals buried in different coffin types with and without bronze funerary vessels at the Xiaohucun site, Henan Province, China.

In addition, past studies developed a specific classification system for Shang Dynasty tombs [3,71], where they can be divided into three different categories (A, B, C) with seven subdivisions (Aa, Ab, Ac, Ba, Bb, Bc, C) (Table 4). Using this system, seven of the tombs (M8, M52, M27, M24, M105, M30, M90) at Xiaohucun are assigned to type Bb. Thus, the owners of these seven tombs are presumed to be junior nobility: six had double coffins and all seven were buried with bronze funerary vessels. Four other tombs (M47, M46, M116, M89) were matched to the Bc category, an indication that the occupants were common people (Table 4). A single tomb (M39) gave somewhat conflicting results. The tomb size was large (3.92 m^2) indicating the owner could have been a noble, but the individual was found to be buried without coffins or bronze funerary objects which is an indication this individual was of the lower class (Table 4). It is important to recognize that these classifications are based on information from the royal Shang capital at Anyang and may not be directly applicable at Xiaohucun. However, their use here provides an important starting point from which to explore the relationship between dietary differences and social status during the late Shang Dynasty.

6.2. Diet and Social Status

No difference is observed between the δ^{13}C results of the perceived elites (n = 7) buried with bronze vessels (−9.1 ± 2.8‰) and the individuals (n = 5) buried without bronze vessels (−8.2 ± 0.7‰), and this indicates that all classes of the population relied heavily on a millet based diet (Figure 5a). However, individuals buried with bronze vessels (10.3 ± 1.6‰) were found to have significantly higher δ^{15}N values (one-way ANOVA; *p* = 0.015) compared to individuals buried without bronze vessels (8.0 ± 0.9‰). This finding suggests that the elite during the late Shang Dynasty were consuming a diet with possibly more animal protein or fish than the commoners. Subtle similar patterns have also been noted at the sites of Xipo (ca. 4000–3300 BC) and Qianzhangda (ca. 1000 BC) [54,56]. In particular, at Qianzhangda, a tomb owner associated with a larger grave was found to have a higher δ^{15}N result compared to the other tomb owners with smaller sized burials, and tomb owners were also found to have higher δ^{15}N results than sacrificed individuals. Zhang et al. [56] concluded that these elevated nitrogen results of the elites with larger tombs were related to increased meat consumption. However, the results presented here are the first to suggest a possible direct isotopic link between diet and social status at a middle Shang Dynasty site from the Central Plains of China.

Table 4. Classification of social status for burials from the Shang and Zhou Dynasties. Note: Information is referenced from IA CASS [3].

Type	Sub-Type	Tomb Passage	Size (m²)	Inner & Outer Coffins	Grave Goods	Human Sacrifice	Owner Status
A	a	4	>100	always	abundant	always	Emperor
	b	2	>20	always	abundant	always	Prince
	c	1	20–70	always	abundant	always	Prince or senior nobility
B	a	0	>10	always	abundant	mostly	Prince or senior nobility
	b	0	3–10	mostly	ordinary	rarely	Junior nobility
	c	0	<3	rarely	little or absence	absence	Common people
C	a	0	N/A	absence	absence	absence	Unprivileged people

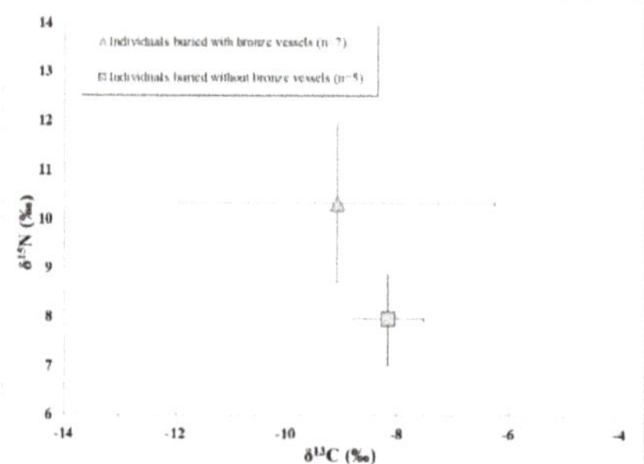

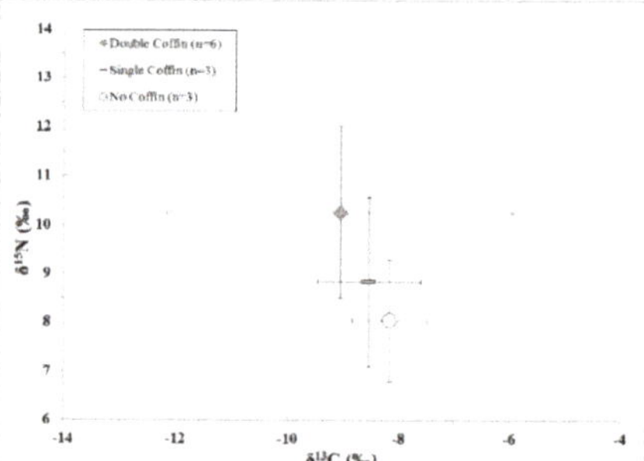

Figure 5. (**a**) δ^{13}C and δ^{15}N results showing the differences between individuals buried with and without bronze vessels at the Xiaohucun site, Henan Province, China. The δ^{15}N results are statistically significant (one-way ANOVA; $p = 0.015$); (**b**) δ^{13}C and δ^{15}N results for individuals buried in a double, single, or without a coffin at the Xiaohucun site, Henan Province, China. The δ^{15}N results were not found to be statistically significant using one-way ANOVA tests (double vs. single, $p = 0.288$; double vs. none, $p = 0.095$).

In addition, the isotopic values of the tombs were examined by coffin type: double coffin, single coffin, and no coffin (Figure 5b). No differences in δ^{13}C were observed for the double (-9.1 ± 3.1‰), single (-8.5 ± 0.9‰), and no coffin (-8.2 ± 0.7‰) burials. Yet again, there were differences in the δ^{15}N values: double (10.2 ± 1.7‰) > single (8.8 ± 1.8‰) > no coffin (8.0 ± 1.3‰) suggesting that possible differences in animal protein consumption were present during the lifetime of these individuals. However, while these results are interesting, they must be viewed with the utmost caution as they were not found to be statistically significant using one-way ANOVA tests (double vs. single coffins, $p = 0.288$; double vs. no coffins, $p = 0.095$).

We further examined the isotopic results of the burials in relation to the number of coffins and tomb sizes since a larger size might equate to higher social status based on the increased labor costs of construction. This permitted an investigation of possible dietary differences within the same social class; specifically did individuals with larger tombs have diets different from individuals with smaller tombs? As with the previous measures of status, the δ^{13}C values show little correlation with the size of the tombs, and the entire population, regardless of status, was focused on the consumption of millet (Figure 6a). The δ^{15}N results of the individuals with a single or no coffin burial show little correlation with tomb size (Figure 6b). In contrast, a significant positive linear correlation ($R^2 = 0.84$) is observed for the $\delta^{15}N$ values between the individuals buried in two coffins and the tomb size. This is possible evidence that owners that were wealthy/powerful enough to build bigger tombs had diets that contained increasing amount of animal products, and the largest tomb (M8) had an occupant that consumed the most animal protein (Figure 6b). While these findings are intriguing and agree with the historical accounts described in Zhou Li "周礼" (Rites of the Zhou Dynasty by Zhou Gong "周公", ca. 1100 BC [88]) and Chunqiu Gongyang Zhuan "春秋公羊传" (Gongyang's Commentary on Spring and Autumn Annals by Gongyang Gao "公羊高", ca. 507 BC [89]) extreme caution is warranted as the sample size is unfortunately small.

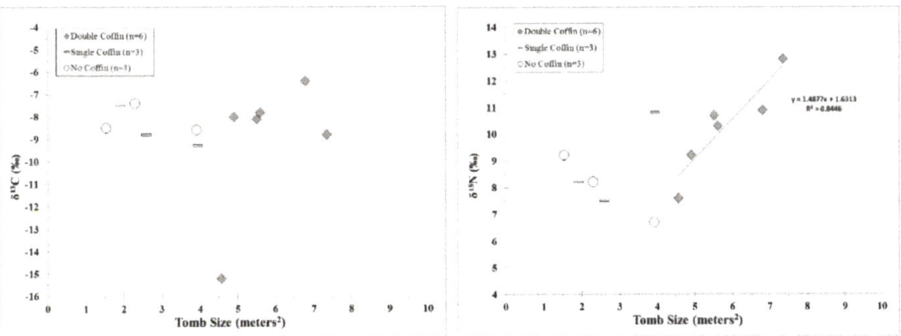

Figure 6. (**a**) $\delta^{13}C$ results vs. tomb size (m^2) for individuals buried in a double, single, or without a coffin at the Xiaohucun site, Henan Province, China; (**b**) $\delta^{15}N$ results vs. tomb size (m^2) for individuals buried in a double, single, or without a coffin at the Xiaohucun site, Henan Province, China.

It is noteworthy that the $\delta^{15}N$ values of some burials did not correlate with the corresponding tomb size although a general relationship between the two variables was observed as mentioned above. The reasons for this are complex and unknown but could be related to the slow turnover rate of bone collagen which averages the general dietary protein isotope signatures of the last decades of life [25,26]. If an individual fell on financial or social misfortune and their social standing was diminished during the later years of their life, this could account for a difference in the social status of their burial vs. Their lifetime diet. For example, the $\delta^{15}N$ result (10.8‰) of M90 was high and he was buried with some bronze funerary objects suggesting a possible elevated social standing during life, but the small size of his tomb (3.95 m^2), shallow burial depth (0.66 m), and the fact that he was only buried in a single coffin could imply that his social status or personal fortunes declined later in life. In addition, the reverse scenario could be envisioned for the M30 burial. This individual had a $\delta^{13}C$ value (-15.2‰) which was radically different that the rest of the population and the lowest $\delta^{15}N$ value (7.6‰) of all the perceived elite burials. This is possible evidence that this person was a commoner that immigrated to the Xiaohucun community, and that this individual was not born into elite status but acquired it later in life, possibly though marriage, accumulation of wealth, or behavior (bravery in battle). However, all of these possibilities are speculative scenarios and addition research is necessary (ancient DNA, sulfur and strontium stable isotope ratios, etc.) to better understand these patterns, and variables such as age and sex should also be considered in future studies that examine Shang Dynasty sites for social stratification with stable isotope ratio analysis.

7. Conclusions

Here, we applied isotopic measurements as a direct technique to determine dietary patterns related to social class at the site of Xiaohucun, Henan Province, China. For the most part, the population was found to be consuming a predominately C_4 diet (millets), although a single individual (M30) was found to have a mixed C_4/C_3 diet and could have been an immigrant to the community. No difference was found in the $\delta^{13}C$ results of the individuals buried with bronze vessels ($-9.1 \pm 2.8‰$) and the individuals buried without bronze vessels ($-8.2 \pm 0.7‰$) but significant differences were present in the $\delta^{15}N$ values: individuals buried with bronze vessels ($10.3 \pm 1.6‰$) vs. individuals buried without bronze vessels ($8.0 \pm 0.9‰$). Isotopic results were then compared by the number of burial coffins that an individual had: double, single, or without coffin. No difference was found in the $\delta^{13}C$ values, but variations were observed in the $\delta^{15}N$ values: double ($10.2 \pm 1.7‰$) > single ($8.8 \pm 1.8‰$) > no coffin ($8.0 \pm 1.3‰$), possible evidence of increased animal protein consumption with higher social status. Lastly, isotopic results and status were examined by the number of coffins and tomb size. Again, no correlation was seen with $\delta^{13}C$, but a linear correlation ($R^2 = 0.85$) was found for the $\delta^{15}N$ values of the elites. Thus, additional social stratification could have existed among the elites with owners' wealthy/power enough to build larger tombs and possibly consuming more animal protein in their diets. These preliminary results of this pilot study offer a glimpse of the social hierarchy that existed during the late Shang Dynasty.

Author Contributions: Conceptualization, N.W.; resource, L.J.; Formal analysis, Y.S.; writing—original draft preparation, N.W.; writing—review and editing, N.W., X.J. All authors have read and agreed to the published version of the manuscript.

Funding: This research was funded by the National Natural Science Foundation of China, grant number 41603009, 41771223, The Ministry of education of Humanities and Social Science project, grant number 16YJCZH100 and National Social Science Foundation of China, grant number 18CKG023.

Acknowledgments: The authors thank Hu Yaowu (Fudan University, Shanghai) and Li Xiaoqiang (Institute of Vertebrate Paleontology and Paleoanthropology, CAS, Beijing) for supporting this study. Thanks to Guo Yi (Zhejiang University, Hangzhou) and Chen Xianglong (Institute of Archaeology, CASS, Beijing) for helpful comments. Wang Tingting and Zhang Xinyu (Institute of Vertebrate Paleontology and Paleoanthropology, CAS, Beijing) are thanked for collagen preparation.

Conflicts of Interest: The authors declare no conflict of interest.

References

1. Liu, L. *The Chinese Neolithic: Trajectories to Early States (New Studies in Archaeology)*; Cambridge University Press: Cambridge, UK, 2007.
2. Liu, L.; Chen, X.C. The Archaeology of China: From the Late Paleolithic to the Early Bronze Age; Cambridge University Press: Cambridge, UK, 2012.
3. IA CASS. *Chinese Archaeology: The Volume of Shang and Xia Dynasties*; China Social Sciences Press: Beijing, China, 2003; pp. 98–101. (In Chinese)
4. IA CASS. *Chinese Archaeology: The Volume of Neolithic*; China Social Sciences Press: Beijing, China, 2010; pp. 788–801. (In Chinese)
5. Dai, S. *Li Ji: Book of Rites*; The North Literature and Art Publishing Houses: Harbin, China, 2013; p. 437. (In Chinese)
6. Xun, K. *Xunzi*; Yan Shi Press: Beijing, China, 2011; p. 658. (In Chinese)
7. Wang, W. The social structure of Zhou dynasty reflects by tomb system and comparisons with Shang dynasty. *Sci. Chi.* **2004**, *3*, 54–55. (In Chinese)
8. Binford, L.R. Mortuary Practices: Their Study and Potential. *Mem. Soc. Am. Archaeol.* **1971**, *25*, 6–29. [CrossRef]
9. Binford, L.R. Meaning, Inference, and the Material Record. In *Ranking, Resource, and Exchange*; Cambridge University Press: Cambridge, UK, 1982; pp. 160–163.
10. O'Shea, J.M. Social Configurations and the Archaeological Study of Mortuary Practices: A Case Study. In *The archaeology of Death*; Chapman, R., Kinnes, I., Randsborg, K., Eds.; Cambridge University Press: Cambridge, UK, 1981; pp. 39–52.
11. O'Shea, J.M. Mortuary Custom in Bronze Age of Southeastern Hungary, Diachronic and Synchronic Perspectives. In *Regional Approaches to Mortuary Analysis: Interdisciplinary Contributions to Archaeology*; Lane, A.B., Ed.; Springer: Berlin, Germany, 1995; pp. 125–145.
12. Saxe, A.A. Social Dimensions of Mortuary Practices in a Mesolithic Population from Wadi Halfa, Sudan. Ph.D. Thesis, Department of Anthropology, University of Michigan, Ann Arbor, MA, USA, 1970.

13. Van der Merwe, N.J.; Vogel, J.C. ^{13}C content of human collagen as a measure of prehistoric diet in woodland North America. *Nature* **1978**, *276*, 815–816. [CrossRef] [PubMed]

14. Schoeninger, M.J.; DeNiro, M.J. Nitrogen and carbon isotopic composition of bone collagen from marine and terrestrial animals. *Geochim. Cosmochim. Acta.* **1984**, *48*, 625–639. [CrossRef]

15. Sealy, J.C.; van der Merwe, N.J.; Thorp, J.A.L.; Lanham, J.L. Nitrogen isotopic ecology in southern Africa: Implications for environmental and dietary tracing. *Geochim. Cosmochim. Acta.* **1987**, *51*, 2707–2717. [CrossRef]

16. Katzenberg, M.A.; Weber, A. Stable isotope ecology and palaeodiet in the Lake Baikal region of Siberia. *J. Archaeol. Sci.* **1999**, *26*, 651–659. [CrossRef]

17. Valentin, F.; Bocherens, H.; Gratuze, B.; Sand, C. Dietary patterns during the late prehistoric/historic period in Cikobia Island (Fiji): Insights from stable isotopes and dental pathologies. *J. Archaeol. Sci.* **2006**, *33*, 1396–1410. [CrossRef]

18. Müldner, G.; Richards, M.P. Stable isotope evidence for 1500 years of human diet at the city of York, UK. *Am. J. Phys. Anthropol.* **2007**, *133*, 682–697. [CrossRef]

19. Choy, K.; Jean, O.R.; Fuller, B.T.; Richards, M.P. Isotopic evidence of dietary variations and weaning practices in the Gaya cemetery at Yeanri, Gimhae, South Korea. *Am. J. Phys. Anthropol.* **2010**, *142*, 74–84. [CrossRef]

20. Bourbou, C.; Fuller, B.T.; Garvie-Lok, S.J.; Richards, M.P. Reconstructing the diets of Greek Byzantine populations (6th–15th centuries AD) using carbon and nitrogen stable isotope ratios. *Am. J. Phys. Anthropol.* **2011**, *146*, 569–581. [CrossRef]

21. Commendador, A.S.; Dudgeon, J.V.; Finney, B.P.; Fuller, B.T.; Esh, K.S. Stable isotope (δ^{13}C and δ^{15}N) perspective on human diet on Rapa Nui (Easter Island) c.a. 1400-1900 AD. *Am. J. Phys. Anthropol.* **2013**, *152*, 173–185. [PubMed]

22. Cui, Y.; Song, L.; Wei, D.; Pang, Y.; Wang, N.; Ning, C.; Li, C.; Feng, B.; Tang, W.; Li, H.; et al. Identification of kinship and occupant status in Mongolian noble burials of the Yuan Dynasty through a multidisciplinary approach. *Philos. Trans. R. Soc. Lond. B. Biol. Sci.* **2015**, *370*, 20130378. [CrossRef] [PubMed]

23. Ma, M.M.; Dong, G.H.; Jia, X.; Wang, H.; Cui, Y.F.; Chen, F.H. Dietary shift after 3600 cal yr BP and its influencing factors in northwestern China: Evidence from stable isotopes. *Quat. Sci. Rev.* **2016**, *145*, 57–70. [CrossRef]

24. Chen, X.L.; Fang, Y.M.; Hu, Y.W.; Hou, Y.F.; Lu, P.; Yuan, J.; Song, G.D.; Fuller, B.T.; Richards, M.P. Isotopic Reconstruction of the Late Longshan Period (ca.4200–3900BP) Dietary Complexity before the Onset of State-Level Societies at the Wadian Site. *Int. J. Osteoarchaeol.* **2016**, *26*, 808–817. [CrossRef]

25. Stenhouse, M.J.; Baxter, M.S. The uptake of bomb ^{14}C in humans, in radiocarbon dating. In *Radiocarbon Dating: Proceedings of the Ninth International Conference, Los Angeles and La Jolla*; Berger, R., Suess, H.E., Eds.; University of California Press: Berkeley, CA, USA, 1979; pp. 324–341.

26. Hedges, R.E.M.; Clement, J.G.; Thomas, C.D.L.; O'Connell, T.C. Collagen turnover in the adult femoral mid-shaft: Modeled from anthropogenic radiocarbon tracer measurements. *Am. J. Phys. Anthropol.* **2007**, *133*, 808–816. [CrossRef]

27. Schwarcz, H.P.; Schoeninger, M.J. Stable isotopic analyses in human nutritional ecology. *Yearb. Phys. Anthropol.* **1991**, *34*, 283–321. [CrossRef]

28. Pechenkina, E.A.; Ambrose, S.H.; Ma, X.; Benfer, J.R.A. Reconstructing northern Chinese Neolithic subsistence practices by isotopic analysis. *J. Archaeol. Sci.* **2005**, *32*, 1176–1189. [CrossRef]

29. Hu, Y.; Wang, S.; Luan, F.; Wang, C.; Richards, M.P. Stable isotope analysis of humans from Xiaojingshan site: Implications for understanding the origin of millet agriculture in China. *J. Archaeol. Sci.* **2008**, *35*, 2960–2965. [CrossRef]

30. Barton, L.; Newsome, S.D.; Chen, F.H.; Wang, H.; Guilderson, T.P.; Bettinger, R.L. Agricultural origins and the isotopic identity of domestication in northern China. *Proc. Natl. Acad. Sci. USA* **2009**, *106*, 5523–5528. [CrossRef]

31. Fu, Q.M.; Jin, S.A.; Hu, Y.W.; Ma, Z.; Pan, J.C.; Wang, C.S. Agricultural development and human diets in Gouwan site, Xichuan, Henan. *Chin. Sci. Bull.* **2010**, *55*, 614–620. [CrossRef]

32. Atahan, P.; Dodson, J.; Li, X.; Zhou, X.; Hu, S.; Bertuch, F.; Sun, N. Subsistence and the isotopic signature of herding in the Bronze Age Hexi Corridor, NW Gansu, China. *J. Archaeol. Sci.* **2011**, *38*, 1747–1753. [CrossRef]

33. Lanehart, R.E.; Tykot, R.H.; Underhill, A.P.; Luan, F.; Yu, H.; Fang, H.; Feinman, G.; Nicholas, L. Dietary adaptation during the Longshan period in China: Stable isotope analyses at Liangchengzhen (southeastern Shandong). *J. Archaeol. Sci.* **2011**, *38*, 2171–2181. [CrossRef]

34. Liu, X.; Jones, M.K.; Zhao, Z.; Liu, G.; O'Connell, T.C. The earliest evidence of millet as a staple crop: New light on neolithic foodways in North China. *Am. J. Phys. Anthropol.* **2012**, *149*, 283–290. [CrossRef] [PubMed]

35. Atahan, P.; Dodson, J.; Li, X.; Zhou, X.; Chen, L.; Barry, L.; Bertuch, F. Temporal trends in millet consumption in northern China. *J. Archaeol. Sci.* **2014**, *50*, 171–177. [CrossRef]

36. Hu, Y.W.; Luan, F.S.; Wang, C.S.; Richards, M.P. Preliminary attempt to distinguish the domesticated pigs from wild boars by the methods of carbon and nitrogen stable isotope analysis. *Sci. China. Earth. Sci.* **2009**, *52*, 85–92. [CrossRef]

37. Chen, X.L.; Yuan, J.; Hu, Y.W.; He, N.; Wang, C.S. Animal feeding practice at Taosi site. *Archaeology* **2012**, *9*, 75–82. (In Chinese)

38. Chen, X.L.; Hu, S.M.; Hu, Y.W.; Ma, R.Y.; Wang, W.L.; Lu, P.; Wang, C.S. Raising practices of Neolithic livestock evidenced by stable isotope analysis in the Wei River valley, North China. *Int. J. Osteoarchaeol.* **2014**, *26*, 42–52. [CrossRef]

39. Hou, L.L.; Hu, Y.W.; Zhao, X.P.; Li, S.T.; Wei, D.; Hou, Y.F.; Hu, B.H.; Lv, P.; Li, T.; Song, G.D.; et al. Human subsistence strategy at Liuzhuang site, Henan, China during the proto-Shang culture (~2000–1600 BC) by stable isotopic analysis. *J. Archaeol. Sci.* **2013**, *40*, 2344–2351. [CrossRef]

40. Bocherens, H.; Drucker, D. Trophic level isotopic enrichment of carbon and nitrogen in bone collagen: Case studies from recent and ancient terrestrial ecosystems. *Int. J. Osteoarchaeol.* **2003**, *13*, 46–53. [CrossRef]

41. Reitsema, L.J. Beyond diet reconstruction: Stable isotope applications to human physiology, health, and nutrition. *Am. J. Hum. Biol.* **2013**, *25*, 445–456. [CrossRef]

42. Bol, R.; Pflieger, C. Stable isotope (^{13}C, ^{15}N and ^{34}S) analysis of the hair of modern humans and their domestic animals. *Rapid. Commun. Mass. Spectrom.* **2002**, *16*, 2195–2200. [CrossRef]

43. Petzke, K.J.; Boeing, H.; Metges, C.C. Choice of dietary protein of vegetarians and omnivores is reflected in their hair protein C-13 and N-15 abundance. *Rapid. Commun. Mass. Spectrom.* **2005**, *19*, 1392–1400. [CrossRef]

44. Katzenberg, M.A. Stable isotope analysis: A tool for studying past diet, demography, and life history. In *Biological Anthropology of the Human Skeleton*; Katzenberg, M.A., Saunders, S.R., Eds.; Wiley-Liss: New York, NY, USA, 2000; pp. 305–327.

45. Ambrose, S.H.; Krigbaum, J. Bone chemistry and bioarchaeology. *J. Anthropol. Archaeol.* **2003**, *22*, 193–199. [CrossRef]

46. Lee-Thorp, J.A. On isotopes and old bones. *Archaeometry* **2008**, *50*, 925–950. [CrossRef]

47. Richards, M.P.; Hedges, R.E.M.; Molleson, T.I.; Vogel, J.C. Stable isotope analysis reveals variations in human diet at the Poundbury Camp cemetery site. *J. Archaeol. Sci.* **1998**, *25*, 1247–1252. [CrossRef]

48. Ambrose, S.H.; Buikstra, J.; Krueger, H.W. Status and gender differences in diet at Mound 72, Cahokia, revealed by isotopic analysis of bone. *J. Anthropol. Archaeol.* **2003**, *22*, 217–226. [CrossRef]

49. Kjellström, A.; Jan Storå, J.; Göran Possnert, G.; Linderholm, A. Dietary patterns and social structures in Medieval Sigtuna, Sweden, as reflected in stable isotope values in human skeletal remains. *J. Archaeol. Sci.* **2009**, *36*, 2689–2699. [CrossRef]

50. Yoder, C. Let them eat cake? Status-based differences in diet in medieval Denmark. *J. Archaeol. Sci.* **2012**, *39*, 1183–1193. [CrossRef]

51. Quintelier, K.; Ervynck, A.; Müldner, G.; Van Neer, W.; Richards, M.P.; Fuller, B.T. Isotopic Examination of Links Between Diet, Social Differentiation, and DISH at the Post-Medieval Carmelite Friary of Aalst, Belgium. *Am. J. Phys. Anthropol.* **2014**, *153*, 203–213. [CrossRef]

52. Dong, Y.; Morgan, C.; Chinenov, Y.; Zhou, L.G.; Fang, W.Q.; Ma, X.L.; Pechenkina, K. Shifting diets and the rise of male-biased inequality on the Central Plains of China during Eastern Zhou. *Proc. Natl. Acad. Sci. USA* **2017**, *114*, 932–937. [CrossRef]

53. Twiss, K. The archaeology of food and social diversity. *J. Archaeol. Res.* **2012**, *20*, 357–395. [CrossRef]

54. Henan Provincial Institute of Cultural Relics and Archaeology. *Xipo Cemetery, Lingbao*; Cultural Relics Press: Beijing, China, 2010. (In Chinese)

55. Ling, X.; Chen, L.; Tian, Y.Q.; Li, Y.; Zhao, C.C.; Hu, Y.W. Carbon and Nitrogen Stable Isotopic Analysis on Human Bones from the Qin Tomb of Sun jianantou Site, Fengxiang, Shaanxi Province. *Acta Anthropologica Sinica* **2010**, *29*, 54–61. (In Chinese)

56. Zhang, X.L.; Qiu, S.H.; Zhong, J.; Liang, Z.H. The carbon, nitrogen stable isotope analysis of human bone from Qianzhangda Cemetery. *Archaeology* **2012**, *9*, 83–90. (In Chinese)

57. Zhang, X.L.; Qiu, S.H.; Zhang, J.; Guo, W. The carbon and nitrogen stable isotope analysis of human bones from Duogang cemetery, Xinjiang. *South. Antiq.* **2014**, *3*, 79–91. (In Chinese)
58. Wang, Y.; Nan, P.H.; Wang, X.Y.; Wei, D.; Hu, Y.W.; Wang, C.S. Dietary Differences in Humans with Similar Social Hierarchies: Example from the Niedian Site, Shanxi. *Acta Anthropologica. Sinica* **2014**, *33*, 82–89. (In Chinese)
59. Li, C. *Anyang*; University of Washington Press: Seattle, WA, USA, 1978.
60. Chang, K.C. *Shang Civilization*; Yale University Press: New Haven, CT, USA, 1980.
61. Loewe, M.; Shaughnessy, E.L. *The Cambridge History of Ancient China: From the Origins of Civilization to 221 B.C.*; Cambridge University Press: Cambridge, UK, 1999.
62. Thorp, R.L. *China in the Early Bronze Age: Shang Civilization*; University of Pennsylvania Press: Philadelphia, PA, USA, 2005; p. 320.
63. Zhu, Y. *The Origin, Migration and Development of Shang Clan*; The Commercial Press: Beijing, China, 2007. (In Chinese)
64. Zhang, G.Z. *Shang Civilization*; SDX Joint Publishing Company: Beijing, China, 2013. (In Chinese)
65. Henan Provincial Institute of Cultural Relics and Archaeology. Unearthing bulletin for late Shang remains of Guandimiao ruins in Xingyang City of Henan Province. *Archaeology* **2008**, *7*, 32–46. (In Chinese)
66. Song, Z.H. *The Social Life History of Xia and Shang Dynasty*; China Social Sciences Press: Beijing, China, 1994. (In Chinese)
67. Song, Z.H. *China Diet History: Xia and Shang Dynasty*; Huaxia Press: Beijing, China, 1999. (In Chinese)
68. Song, Z.H. *Social Lives and Rituals in Shang Dynasty*; China Social Sciences Press: Beijing, China, 2010. (In Chinese)
69. Wang, X. Dietetic rules and regulations research in Zhouli. Master's Thesis, Yangzhou University, Yangzhou, China, 2007. (In Chinese).
70. Wang, Y.X.; Xu, Y.H. *Countries and Society in Shang Dynasty*; China Social Sciences Press: Beijing, China, 2011. (In Chinese)
71. Tang, J.G. The Social Organization of Late Shang China: A Mortuary Perspective. Ph.D. Thesis, University of London, London, UK, 2004.
72. Yu, W.C.; Gao, M. The study of the system with the tripod at Zhou dynasty. *J. Peking Univ. Philosophy Soc. Sci.* **1978**, *2*, 84–97. (In Chinese)
73. Reinhart, K. Ritual feasting and empowerment at Yanshi Shangcheng. *J. Anthropol. Archaeol.* **2015**, *39*, 76–109. [CrossRef]
74. Qing, L. Type value and grave value: An introduction to a quantifying method in tomb research. *Huaxia Archaeol.* **2007**, *3*, 133–137. (In Chinese)
75. Wu, F. The research of funeral furniture of Shang Dynasty. Master's Thesis, Zhengzhou University, Zhengzhou, China, 2012. (In Chinese).
76. Jia, L.M.; Zeng, X.; Liang, F.; Yu, H.W. The Xiaohucun late Shang noble cemetery of Xiangyang city, Hennan. In *Archaeological Society of China, Important Archaeology Discoveries in China of 2006*; Cultural Relics Press: Beijing, China, 2007. (In Chinese)
77. Lee, G.; Crawford, G.W.; Liu, L.; Chen, X. Plants and people from the early Neolithic to Shang periods in North China. *Proc. Natl. Acad. Sci. USA* **2007**, *104*, 1087–1092. [CrossRef]
78. Qiu, S.H.; Zhong, J.; Zhao, X.P.; Sun, F.X.; Cheng, L.Q.; Guo, Y.Q.; Li, X.W.; Ma, X.L. Studies on Diet of the Ancient People of the Yangshao Cultural Sites in the Central Plains. *Acta Anthropol. Sin.* **2010**, *29*, 197–207. (In Chinese)
79. IA CASS. *Science for Archaeology: The Third Volume*; Science Press: Beijing, China, 2011; pp. 1–35. (In Chinese)
80. Liu, X.; Lightfoot, E.; O'Connell, T.C.; Wang, H.; Li, S.; Zhou, L.; Hu, Y.; Motuzaite-Matuzeviciute, G.; Jones, M.K. From necessity to choice: Dietary revolutions in west China in the second millennium BC. *World Archaeol.* **2014**, *46*, 661–680. [CrossRef]
81. Jia, L.M.; Wang, W.B.; Lu, H.W.; Liang, F.W.; Zeng, X.M. Archaeological Excavation of Xiaohucun village cemetery in Xingyang, Henan Province. *Huaxia Archaeol.* **2015**, *1*, 3–25. (In Chinese)
82. Zhu, H. *Physical Anthropology*; Higher Education Press: Beijing, China, 2004. (In Chinese)
83. Richards, M.P.; Hedges, R.E.M. Stable isotope evidence for similarities in the types of marine foods used by Late Mesolithic humans at sites along the Atlantic coast of Europe. *J. Archaeol. Sci.* **1999**, *26*, 717–722. [CrossRef]

84. DeNiro, M.J. Postmortem preservation and alteration of in vivo bone collagen isotope ratios in relation to palaeodietary reconstruction. *Nature* **1985**, *317*, 806–809. [CrossRef]

85. Zhang, X.L.; Wang, J.X.; Xian, Z.Q.; Qiu, S.H. Studies on ancient human diet. *Archaeology* **2003**, *2*, 62–75. (In Chinese)

86. Dai, L.; Li, Z.; Zhao, C.; Yuan, J.; Hou, L.; Wang, C.; Fuller, B.T.; Hu, Y. An isotopic perspective on animal husbandry at the Xinzhai site during the initial stages of the legendary Xia Dynasty (2070—1600 BC). *Int. J. Osteoarchaeol.* **2015**. [CrossRef]

87. IA CASS. *Science for Archaeology: The Second Volume*; Science Press: Beijing, China, 2007; pp. 1–34. (In Chinese)

88. Xu, Z.Y.; Chang, P.Y. *Zhou Li: Translation and Annotation*; Zhonghua Book Company: Shanghai, China, 2014. (In Chinese)

89. Liu, S.C. *Chunqiu Gongyang Zhuan: Translation and Annotation*; Zhonghua Book Company: Shanghai, China, 2011. (In Chinese)

Article

Climate Change and the Pattern of the Hot Spots of War in Ancient China

Shengda Zhang [1], David Dian Zhang [1,*] and Jinbao Li [2]

[1]   School of Geographical Sciences, Guangzhou University, Guangzhou 510006, China
[2]   Department of Geography, The University of Hong Kong, Hong Kong, China
*   Correspondence: dzhang@gzhu.edu.cn

Received: 15 March 2020; Accepted: 11 April 2020; Published: 13 April 2020

Abstract: Quantitative research on climate change and war hot spots throughout history is lacking. In this study, the spatial distribution and dynamic process of war hot spots under different climatic phases in imperial China (1–1911 CE) are revealed using Emerging Hot Spot Analysis (EHSA), based on the Global Moran's Index for testing the degree of spatial autocorrelation or dependency. The results show that: (1) Battles were significantly clustered regardless of any climatic mode or war category. (2) Hot spots for all war were generally located in the Loess Plateau and the North China Plain during warm and wet periods, but in the Central Plain, the Jianghuai region, and the lower reaches of the Yangtze River/Yangtze River Delta during cold and dry conditions. (3) Hot spots for agri-nomadic conflict have similar patterns as those for all war, whereas rebellion hot spots expanded outward during warm and wet intervals yet contracted inward during cold and dry stages. These findings, by providing insightful evidence into the spatiotemporal patterns of war under the movements of climatic-ecological zones and geopolitical variations in ancient China, can be a starting point for future exploration of the long-term relationship between climate change and social security.

Keywords: climate change; war; imperial China; Global Moran's *I*; Emerging Hot Spot Analysis

---

1. Introduction

The climate–war nexus in historical China has been widely addressed by academic communities over the past decade [1–19]. Scientists have mostly focused on this nexus from a temporal or time-series angle, whereas the spatiality of war and its connection with climate change has rarely been investigated. Recently, our group determined that in imperial China, (1) geopolitical variables, such as the boundary between agriculturalists and pastoralists, size of agricultural empire, battle location, and the direction of war, were affected by multi-centennial precipitation fluctuation [15]; (2) the distributions of natural disasters (flood and drought) and their social impacts (famine, cannibalism, and war) were influenced by population on provincial and decadal scales [20]; and (3) secular temperature variation fundamentally regulated the spatial disparities of war via controlling agricultural and pastoral productivity [21]. However, research on the spatiotemporal pattern and its dynamic process of war has not been fully conducted.

In this study, we aimed to solve this problem by examining the linkage between climate change and the focus (or hot spot) of war in imperial China. Using the comprehensive official history and well-preserved local and private records in China since ancient times, a few native scholars have discovered the focus (similarly, geographical pivot, or strategic area) of war. For example, Song [22] divided wars in the imperial era into frontier and interior wars, and the pivots of the latter were distributed in the western Henan Corridor, the south of the Huai River (also known as the Jianghuai region), and Jing–Xiang (present-day northern Hubei). Rao [23] introduced an irregular chessboard pattern of war from a military geographical view, in which there were nine strategic areas—four corners: Guanzhong, Hebei, Southeast China, and Sichuan; four foci/pivots: Shanxi, Shandong, Hubei, and Hanzhong; as well as the heartland, the Central Plain. Wang [24] extracted information on the spatial distribution and the shift of the focus area of war from poems in 618–765 CE (i.e., the early Tang dynasty), which implied the potential value of classic literature. Leng [25] looked into the frontier conflicts between the central governments and northern minorities during the imperial age and found that the focus areas had moved from the Hetao region and the Hexi Corridor of Northwest China since the Qin and Han dynasties to the Sixteen Prefectures of Yanyun and the western Liaoning Corridor of Northeast China after the Tang dynasty. These findings, however, are all qualitative and do not contain any climatic variables.

To fill in the research gaps, the technique in ArcMap, Emerging Hot Spot Analysis (EHSA), was applied in this study. Developed by the Environmental Systems Research Institute, Inc. (ESRI), EHSA identifies the spatial trends and distributions of different types of hot spots from data points. It has been employed by some experts to unveil spatial patterns with time, such as detecting public sentiment from geotagged photo collections in San Francisco in 2006–2015 and showing that different emotions (anger, disgust, fear, joy, sadness, and surprise) have distinct spatial distributions [26], as well as the spatiotemporal associations between a community greening program and neighborhood crime rates in Flint, Michigan, in 2005–2014 [27]. Other examples include the spatiotemporal analysis of changes in lode mining claims around the McDermitt Caldera, northern Nevada, and southern Oregon, in 1976–2010 [28], the expected trend in the occurrence of pulmonary tuberculosis cases from Hamadan Province, Iran, during 2005–2013 [29], spatial patterns of crimes (larceny and aggravated assault) in Miami-Dade County, Florida, from 2007 to 2015 [30], and statistically significant temporal-spatial trends of forest loss in Brazil, Indonesia, and the Democratic Republic of Congo between 2000 and 2014 [31]. Thus, by integrating time and space domains, EHSA was suitable for the task of uncovering the hot spots of war in China from 1 to 1911 CE. Compared with the aforementioned studies that only covered several decades at most, this work is the first to use EHSA on a long time scale. Furthermore, we made a methodological breakthrough by using the analysis with a climatological background, which may lay a foundation for further exploration by researchers in related fields.

The structure of this paper is as follows. Data sources and data processing, which include the cyclic division schemes for temperature and precipitation series, and the statistical tools, such as the Global Moran's *I* and EHSA, are introduced in Section 2. Distributions of the hot spots of three kinds of war—all war, the conflict between agriculturalists and pastoralists ("agri-nomadic conflict"), and rebellion under different climatic phases (warm versus cold/wet versus dry) in ancient China are presented in Section 3. Some discussions about the effects of climatic and other non-climatic factors on war hot spots, and our conclusions, are provided in Sections 4 and 5, respectively.

2. Materials and Methods

2.1. Data Source and Data Processing

The data used in this study included climatic series and battle coordinates. The procedures of cycle divisions for temperature and precipitation sequences are stated in detail.

2.1.1. Climate

Derived from Ge et al. [32], the paleotemperature anomalies were reconstructed with the partial least squares regression method based on multi-proxies (lake sediments, stalagmites, historical documents, tree rings, and ice cores) from five regions (northeast, central east, southeast, northwest, and the Tibetan Plateau) of China during 5–1995 CE. Due to the relatively coarse decadal resolution, the data were linearly interpolated into an annual sequence. The original data were smoothed with a five-point fast Fourier transform (FFT) filter to represent the 50-year variation. Similarly, in this study, the FFT filter was set to 50 points (via OriginLab 2018) to obtain the same low-frequency signal.

The paleoprecipitation series during 300 BCE–2000 CE was synthesized by Zhang et al. [15] from 38 document-based single-proxy hydroclimatic datasets at an annual resolution, using the composite plus scale method [33]. This is the first-published long precipitation sequence that spans the past 2000 years on the national scale. A 300-year Butterworth low-pass filter was applied to retrieve the multi-centennial cycle [15]. In this study, the imperial age 1–1911 CE was taken for consistency, and the average of this sequence was calculated to facilitate comparisons.

2.1.2. Division of Climatic Cycles

On the basis of a few criteria, the temperature and precipitation series were divided into different warm–cold and wet–dry cycles.

Temperature: Seven Warm–Cold Cycles

The paleotemperature anomalies reconstructed by Ge et al. [32] were originally divided into four warm (5–200, 551–760, 951–1320, and 1921 CE–present) and cold (201–350, 441–530, 781–950, and 1321–1920 CE) intervals. These do not cover the entire period, as this division scheme does not indicate whether 351–440, 531–550, and 761–780 CE belonged to warm or cold periods. In this study, by referencing the definitions of warm and cold phases from Zhang et al.'s [1] first study on the relationship between climate change, social unrest, and dynastic transition in historical China, the start and end of a warm or cold stage were placed in the midpoint between the highest and lowest temperature anomalies. Based on this criterion, the procedure of cyclic division was as described below:

(1) Determination of the maxima and minima over the temperature sequence: The maximum in 575 CE and minimum in 615 CE were excluded due to the short duration of the resultant warm (530–594 CE) and cold (595–659 CE) periods that lasted for only 65 years. The maximum in 1911 CE was disregarded, as the warm phase in 1873–1911 CE would have been too short to avoid any possible bias otherwise. Another minimum in 95 CE and maximum in 115 CE had a slight difference in temperature anomalies and were not considered. Accordingly, seven pairs of extrema were identified during 5–1911 CE.

(2) Calculation of the midpoint for each pair of extrema: The midpoint from a maximum to a minimum was set as the end of a warm stage, and the year after was treated as the start of a cold phase. By contrast, the midpoint from a minimum to a maximum was set as the end of a cold period, and the year after was treated as the start of a warm stage. The maximum of 1911 CE was excluded in the calculations.

Therefore, seven temperature cycles were defined according to this division scheme (Figure 1a). Each warm and cold phase is listed in Table 1. The duration of all warm periods was 916 years, while cold periods covered 991 years in total. Hence, the warm and cold intervals were relatively balanced compared with the results from Ge et al. [32], in which there were 855 and 1010 years of warm and cold stages, respectively.

Precipitation: Three Wet–Dry Cycles

The paleoprecipitation reconstruction from Zhang et al. [15], which was used to discuss the interactions between agriculturalists and pastoralists in imperial China, was divided into three "Yang" (agriculturalist empires took control of the hinterland) and "Yin" (nomadic tribes invaded and established their own regimes on agricultural region) periods. The Yang–Yin division scheme is actually dynasty-oriented rather than climate-based. Specifically, Yin 1 happened from the Eastern Jin dynasty in 317 CE until the end of the Southern and Northern dynasties in 589 CE, Yin 2 occurred during the Southern Song–Yuan dynasty (1127–1368 CE), and Yin 3 coincided with the Qing dynasty (1644–1911 CE). Here, a new criterion for re-delimiting the precipitation curve was established, in which a wet (dry) phase can be defined when the 300-year smoothed precipitation is above (below) the average of the series (about 666.7 mm/year). Hence, three precipitation cycles were determined, and over the study period, the lengths of all wet and dry periods lasted for 826 and 1085 years, respectively (Figure 1b and Table 1).

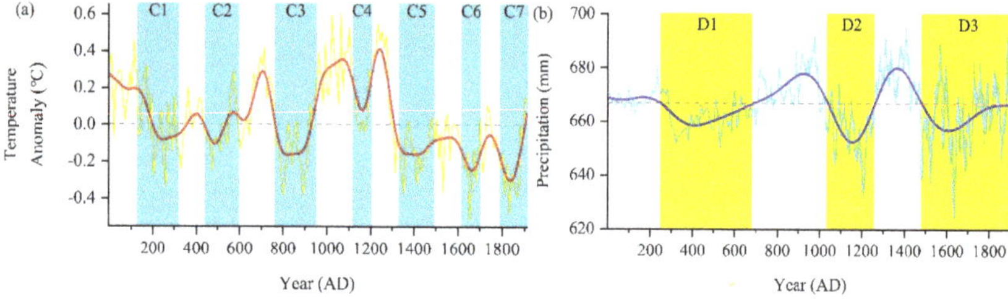

Figure 1. (**a**) Paleotemperature anomaly under the seven-cycle scheme. Note: The annual and 50-year FFT-smoothed series are colored in yellow and red, respectively. Seven cold phases are highlighted in cyan bars. (**b**) Paleoprecipitation reconstruction under the three-cycle scheme. Note: The annual and 300-year curves (smoothed by Butterworth low-pass filter) are shown in cyan and blue, respectively. Three dry intervals are marked by yellow bars.

Table 1. Divisions of warm (W)/cold (C) and wet (W)/dry (D) cycles in ancient China, 1–1911 CE.

Warm Period	Year	Cold Period	Year	Wet Period	Year	Dry Period	Year
W1	5–125	C1	126–320	W1	1–249	D1	250–685
W2	321–440	C2	441–595				
W3	596–765	C3	766–950	W2	686–1041	D2	1042–1262
W4	951–1120	C4	1121–1205				
W5	1206–1330	C5	1331–1490				
W6	1491–1615	C6	1616–1705	W3	1263–1483	D3	1484–1911
W7	1706–1790	C7	1791–1911				

2.1.3. War

War data were gathered from the Tabulation of Wars in Ancient China, an appendix that belongs to the Military History of China, which was summarized by the Editorial Committee of Chinese Military History [34] and has been extensively employed in previous research [1–5,8–10,14–20]. In this study, battle was considered the basic unit of war. The criterion is that if two sides engaging in a war have a fight in reality, then such a fight is regarded as a battle. The terms 'battle' and 'war' are interchangeable in this study when referring to different categories. All the ancient battlefields in this compendium were verified using the Historical Atlas of China [35], counted within the present territory of China, and converted into currently used place names. In other words, battlefields beyond the national boundary were excluded even though they were historically in the areas that belonged to the Chinese Empire. The exclusion of such battlefields may affect the spatial pattern of war, yet these outliers only account for a small proportion of all records—they are usually difficult to locate due to extreme remoteness and the lack of documents. Thus, 5501 battlefields during 1–1911 CE were identified in this study. This number only represents battle locations with definite coordinates, which means the actual number is larger. However, battles do not actually occur at an exact point. From a micro perspective, a battle should have a combat zone, which cannot be measured because people never know much about it. Therefore, the hot spots derived from battle points or coordinates can only be examined in a large framework. The spatial scale of this study was set to national rather than regional and local scales. The definitions of different kinds of wars (i.e., agri-nomadic conflict and rebellion) are provided in Supplementary Materials.

2.2. Methods

In the "Spatial Autocorrelation" tool of ArcMap, the Global Moran's *I* statistic, based on feature locations and attribute values [36], was adopted to measure the degree of global spatial autocorrelation (or, roughly speaking, to decide whether there was a spatial cluster effect) for different kinds of war in historical China. Before the tool can be applied, a grid containing the point data of battlefields was generated to facilitate the statistical analysis. In this study, 100 km was selected as the length of each square cell (Figure 2, and the reason for this is detailed in Supplementary Materials). Then, the number of battles in each cell of the grid was counted. Next, the Global Moran's *I* statistic was calculated according to

$$I = \frac{n}{S_0} \frac{\sum_{i=1}^{n} \sum_{j=1}^{n} w_{i \cdot j}(x_i - \bar{x})(x_j - \bar{x})}{\sum_{i=1}^{n}(x_i - \bar{x})^2} \tag{1}$$

where n is the total number of cells, x_i and x_j are the counted battle number in cells i and j respectively, $\bar{x}$ is the mean value of battle number in all cells, and $w_{i,j}$ denotes the proximity between i and j. When i and j are adjacent, $w_{i,j} = 1$; otherwise, $w_{i,j} = 0$. S_0 denotes the aggregate of all spatial proximity as

$$S_0 = \sum_{i=1}^{n} \sum_{j=1}^{n} w_{i,j} \tag{2}$$

The z-score for the Global Moran's *I* statistic is computed as

$$z = \frac{I - E[I]}{\sqrt{V[I]}} \tag{3}$$

where $E[I]$ and $V[I]$ are the expectation and variance, respectively, with the formulas

$$E[I] = -1/(n - 1) \tag{4}$$

$$V[I] = E[I^2] - E[I]^2 \tag{5}$$

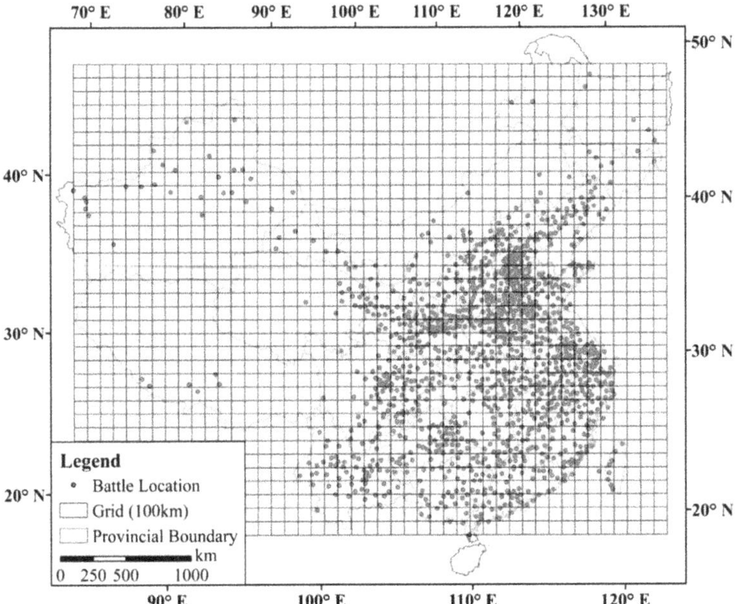

Figure 2. Battle location and the grid (100 × 100 km per cell) adopted in this study.

The index ranges between −1 and 1. The spatial autocorrelation is positive (negative) if the value is larger (smaller) than 0. The higher (lower) the value, the more clustered (dispersed) the war. There is no spatial autocorrelation (i.e., random distribution) when the value is equal to 0.

As the Global Moran's I becomes positive, it is possible that there are some hot spots in which battles are clustered in space. Thus, in this research, EHSA was used to visualize the foci of war. The operational procedure of EHSA is realized in two major steps. First, a space–time cube with network common data form (.nc format) was created by running the tool "Create Space Time Cube" in the package "Space Time Pattern Mining Tools" of ArcMap. Instead of real 'points' (e.g., battle coordinates), the hot spots are an array of grids in which point data are aggregated and counted. Each cell (or bin) must have the same size. In this study, 100 km was chosen as the length of the side for each bin, because a bin less than 100 km (e.g., 50 km) would lead to innumerable cells, each of which only covers a small area. In this case, the bin number that contains battlefields may be small, whereas others may be largely blank. By contrast, if the length of the side is more than 100 km (e.g., 200 km), the resultant patterns may be too coarse to decipher, since only limited amounts of bins reside in the study area (Figure S1). Therefore, the medium size of 100 × 100 km was suitable for this task.

During the first step, the Mann–Kendall trend test (or M–K test) was automatically conducted to evaluate the overall trend for all bins in the cube. It is a non-parametric test used to analyze whether data are consistently increasing or decreasing over time [37,38]. More details about the M–K test are provided in Supplementary Materials. Second, the cube was input in the Emerging Hot Spot Analysis tool, and the Getis-Ord Gi^* statistic (i.e., the traditional hot spot analysis) was calculated for each bin. The formula for obtaining the Gi^* is given as [39]

$$G_i^* = \frac{\sum_{j=1}^n w_{i,j} x_j - \overline{x} \sum_{j=1}^n w_{i,j}}{S \sqrt{\frac{n \sum_{j=1}^n w_{i,j}^2 - \left(\sum_{j=1}^n w_{i,j}\right)^2}{n-1}}} \tag{6}$$

where

$$\overline{x} = \frac{\sum_{j=1}^n x_j}{n} \tag{7}$$

$$S = \sqrt{\frac{\sum_{j=1}^n x_j^2}{n} - (\overline{x})^2} \tag{8}$$

Once the analysis was completed, each bin had an associated z-score, p-value, and hot spot classification. With the resultant trend, z-score, as well as p-value for each location, the EHSA tool categorized eight kinds of hot spots, which are listed in Table 2.

Table 2. Definitions of various hot spots in EHSA.

Category	Definition
New	The most recent time step interval is hot for the first time.
Consecutive	A single uninterrupted run of hot time step intervals, being comprised of less than 90% of all intervals.
Intensifying	At least 90% of the time step intervals are hot and are becoming hotter through time.
Persistent	At least 90% of the time step intervals are hot, with no trend up or down.
Diminishing	At least 90% of the time step intervals are hot and are becoming less hot over time.
Sporadic	Some of the time step intervals are hot.
Oscillating	Some of the time step intervals are hot, some are cold. The most recent time step interval is hot.
Historical	At least 90% of the time step intervals are hot, but the most recent time step interval is not.

(http://desktop.arcgis.com/en/arcmap/latest/tools/space-time-pattern-mining-toolbox/learnmoreemerging.htm).

In terms of the spatiality of war, EHSA surpasses the traditional hot spot analysis because bins with high frequencies of battles surrounded by high values can be determined and various hot spots are categorized based on the variation trends over time. Accompanying hot spots, however, the spatiotemporal patterns of cold spots (i.e., clusters of low values) are generated, which may be meaningless since war hot spots were prioritized in this study. Only significant hot spots are presented ($p < 0.05$).

3. Results

3.1. Global Moran's I

The results of the Global Moran's *I* are provided in Table 3. The values ranged between 0.37 and 0.59 and were all significantly positive ($p < 0.01$), which indicates the cluster effect of war (i.e., positive spatial autocorrelation or spatial dependency) and the feasibility of EHSA. In addition, except for agri-nomadic conflict between warm and cold phases, the statistics during all cold and dry intervals were larger than those in warm and wet stages. This illustrates that battles became more concentrated in cold and dry climate, but were slightly scattered in warm and wet conditions. The indexes for all war at any climatic mode were the largest, probably because each cell of the grid contains more battlefields. This is followed by rebellion, and agri-nomadic conflict had the smallest values (except that the value for rebellion in warm periods was the smallest), which means there seems to be a positive correlation between battle number and the degree of the concentration of war.

Table 3. The Global Moran's *I* for three kinds of war in all warm/cold and wet/dry intervals.

Climatic Phase	All War	Agri-Nomadic Conflict	Rebellion
Warm	0.508 **	0.471 **	0.422 **
Cold	0.588 **	0.416 **	0.572 **
Wet	0.487 **	0.373 **	0.457 **
Dry	0.579 **	0.491 **	0.542 **

Note: ** $p < 0.01$.

3.2. Mann–Kendall Trend Test

Table 4 shows that except for wet periods, the hot spots for all war during the other climatic phases failed to pass the M–K test as the trend statistics were insignificant. The statistics for agri-nomadic conflict at any climatic stage were significantly positive compared with those for all war. The value for all warm intervals was the largest (8.446, $p < 0.01$) amongst all statistics. For rebellion, the counterpart of all warm stages failed to pass the M–K test at the level of 0.05, whereas other statistics were positively significant, and the value for wet conditions was the largest (4.075, $p < 0.01$). Thus, the results of the trend test indicate that notwithstanding the difference in climatic phase, the frequencies of agri-nomadic conflict and rebellion in imperial China basically increased through time.

Table 4. Trend statistic for three categories of war during all warm/cold and wet/dry phases

Climatic Phase	All War	Agri-Nomadic Conflict	Rebellion
Warm	0.979	8.446 **	1.548
Cold	−0.596	3.130 **	2.207 *
Wet	4.445 **	3.103 **	4.075 **
Dry	−0.775	6.678 **	3.026 **

Note: ** $p < 0.01$, * $p < 0.05$.

3.3. EHSA Pattern and Explanation

The explanations of the spatial patterns of EHSA involve many geographical names (Figure S2) and historical periods (Table S1) in China. Figure 3a shows that hot spots were preponderantly distributed in northern China, i.e., from the border between Qinghai and Gansu to western Liaoning, during all warm stages. Only a few oscillating hot spots were located in the Yangtze River Delta. Intensifying hot spots indicate that the areas were becoming increasingly hot (i.e., battles were becoming increasingly frequent). They were chiefly concentrated in central Shaanxi–eastern Gansu and southern

Shanxi–the North China Plain and surrounded by other types. Beyond that, the south and north were predominantly occupied by historical versus sporadic and oscillating hot spots, respectively. As historical hot spots are hot most of the time but not hot in the most recent time, this situation implies the alteration of hot spots in history. To the north, probably along the Great Wall, battles were not as frequent as those in intensifying hot spots; therefore, they were categorized as sporadic or oscillating. Similarly, battles were occasionally clustered, but scattered or even absent sometimes in the Yangtze River Delta, which created oscillating hot spots.

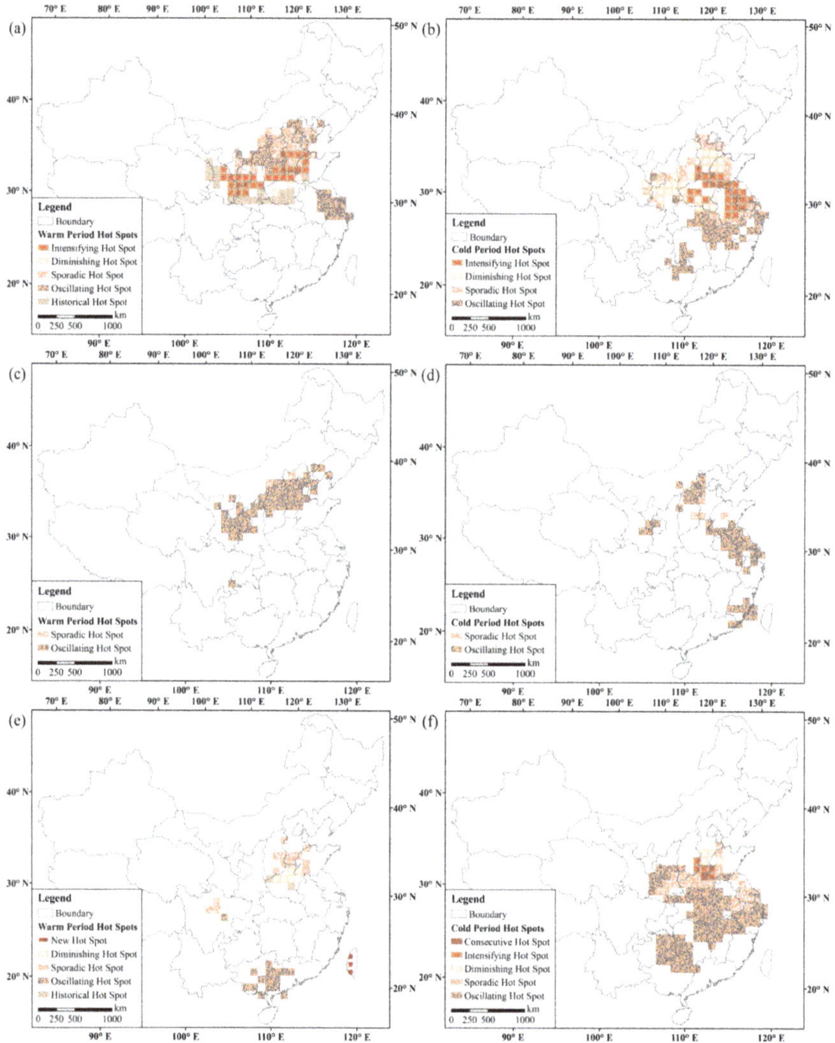

Figure 3. EHSA patterns of all warm and cold periods for three types of war: (**a,b**) All war, (**c,d**) agri-nomadic conflict, and (**e,f**) rebellion. Only significant hot spots ($p < 0.05$) are shown (the same as Figure 4).

By contrast, intensifying hot spots moved massively southeastward to the Central Plain and more prominently to the Jianghuai region and the lower reaches of the Yangtze River during all cold intervals (Figure 3b). The original hot spots in warm periods turned into diminishing hot spots, along with a few sporadic or oscillating cells in central Shaanxi–eastern Gansu and the North China Plain. Since diminishing hot spots are opposite to intensifying hot spots in nature, this alteration indicates a less hot pattern of war (or less frequent battles) in northern China over time. Furthermore, the hot spots in all cold stages were generally situated farther south/southeast as they disappeared from northeastern Qinghai and central Gansu to the northern Loess Plateau and northern Hebei, whereas they appeared to the south of 30° N in a larger proportion. However, as battles in the middle–lower reaches of the Yangtze River–Yangtze River Delta and near the border of Hunan, Guangxi, and Guizhou were only concentrated during certain intervals, the hot spots in these areas were categorized as oscillating or sporadic.

Figure 3c,d visualize the EHSA patterns for agri-nomadic conflict over all warm and cold stages. In Figure 3c, there were only two sporadic hot spots, while others were oscillating hot spots, extending from central Gansu to western Liaoning. This pattern represents the confrontations between agricultural empires and nomadic tribes/regimes along the Great Wall in historical China. However, when carefully examining the time-series, the conflicts contained by oscillating hot spots were concentrated in two main periods: The Northern Song and Ming dynasties (Table S1). The war against the Liao and Western Xia for the former, and that against the Mongols for the latter, occupied the largest proportion, whereas battles during other stages were not intensive. This result is in line with the definition of oscillating hot spot, i.e., occasionally hot (highly clustered), occasionally cold (sparsely scattered), and the last time step (i.e., in the Ming dynasty) is hot.

In comparison, the hot spots for agri-nomadic conflict in cold climate principally shifted southeastward (Figure 3d). Although some cells remained in northern China, they do not match the scale in warm stages. Instead, hot spots (again prevailingly classified as oscillating) were more concentrated in the Jianghuai region, the lower reaches of the Yangtze River–Yangtze River Delta, and Fujian. Similarly, they are related to the distributions of battles during certain phases. Those in the Jianghuai region and the lower reaches of the Yangtze River–Yangtze River Delta resulted from N–S confrontations, such as northern dynasties versus southern dynasties and Jin versus Southern Song (Table S2). The hot spots in Fujian could be explained by the Ming–Qing war. Given that these oscillating hot spots emerged in different periods, the existence of this category is understandable.

The EHSA patterns for rebellion in all warm and cold stages are depicted in Figure 3e,f. In Figure 3e, the hot spots in all warm periods expanded outward and were clustered in four separate parts: Sichuan, the Central Plain, Taiwan, and Guangxi–Guangdong. Five kinds of hot spots were generated: new, diminishing, sporadic, oscillating, and historical. Historical and diminishing hot spots only emerged in the Central Plain, where rebellions were concentrated in earlier eras but gradually became less frequent. Oscillating hot spots mostly appeared in Guangxi and Guangdong, in which battles largely occurred in later eras but were not dense. Sporadic hot spots were distributed in the Chengdu Plain and its northwest, with a few adjacent to diminishing hot spots in northern China. Battles in these areas were concentrated, yet they were occasionally intensive in some intervals. Finally, the new hot spots in western Taiwan may denote the recent revolts during the Qing dynasty.

The rebellion hot spots in all cold phases were more clustered (Figure 3f). The majority of them were oscillating hot spots, spreading from eastern Guizhou to northern Zhejiang and occupying the middle reaches of the Yangtze River. Another group in the Loess Plateau was adjacent to sporadic hot spots, which extended from Guanzhong to the Central Plain, with another part emerging in the Yangtze River Delta. Other types—such as consecutive, intensifying, and diminishing hot spots—appeared in northern China with a few cells only. Diminishing hot spots were situated north of intensifying hot spots, which implies the southward movement of the war focus through time. A comparison with the warm climate pattern showed that the hot spots in the northern Central Plain (southern North China Plain) changed from diminishing (sporadic) cells in all warm intervals to intensifying–consecutive (diminishing) ones in all cold phases. As a traditional warring zone in ancient China, the northern Central Plain became increasingly hot in recent time steps, particularly during the Ming–Qing transition and the late Qing dynasty. Sporadic hot spots in the Guanzhong–Central Plain and the Yangtze River

Delta were not always hot. To the south, oscillating hot spots primarily included the rebellions during later cold stages, indicating the inward contraction of rebellion in cold periods.

The spatial patterns derived from EHSA during all wet and dry stages are visualized in Figure 4. Figure 4a presents the result of all wet periods for all war. Similar to Figure 3a, four types of hot spots (intensifying, diminishing, sporadic, and oscillating) dominated northern China and stretched from eastern Gansu to the North China Plain via Guanzhong, Shanxi, and the Central Plain. They also spread southeastward to the Jianghuai region, the lower reaches of the Yangtze River, and the Yangtze River Delta, but were exclusively covered by oscillating hot spots. Intensifying hot spots were clustered in most of the North China Plain, where battles became increasingly frequent through time. By contrast, in Guanzhong and the eastern Central Plain, the cells that belonged to diminishing hot spots became gradually less hot or not hot in the end, which implies a possible northward movement of battle over time. A striking contrast was observed between the patterns of all wet and dry stages. In Figure 4b, intensifying hot spots shifted to present-day Jiangsu, while the cells in the north mostly turned into diminishing hot spots. Others, such as sporadic and oscillating hot spots, laid in the north (along 40° N) and south (the lower reaches of the Yangtze River) respectively, with a few historical cells scattered around.

Figure 4c,d present the EHSA patterns for agri-nomadic conflicts in all wet and dry stages. The hot spots during wet periods were all clustered in northern China, i.e., from central Ningxia to eastern Hebei, and included three types: intensifying, sporadic, and oscillating. Oscillating hot spots were mainly distributed in Ningxia, northeastern Gansu, and northern Shaanxi, while intensifying hot spots in the North China Plain were surrounded by sporadic ones. The differences among them may have resulted from the proportions of time steps in three wet phases. Still, most hot spots were found in the north during all dry stages (Figure 4d), but their range elongated outward from two sides. The western end extended into central Gansu, while the eastern end extended into northern Liaoning. A few hot spots appeared in the lower reaches of the Yangtze River, which differentiates the patterns between wet and dry phases. All hot spots in northern China were oscillating, indicating the confrontations between agricultural and nomadic regimes along the Great Wall, which escalated in later times. Sporadic and new hot spots emerged in the lower reaches of the Yangtze River and coastal Fujian, respectively. Hence, the hot spots in southern China represent the southward nomadic invasions, and the new hot spot can be associated with the Manchu conquest during the Ming–Qing transitional period.

The EHSA results for rebellion in all wet and dry intervals are displayed in Figure 4e,f, which are similar to the patterns for all warm and cold stages in Figure 3e,f. The three separate parts in all wet phases indicate that hot spots were clustered in the Central–North China Plain, the lower reaches of the Yangtze River, as well as Guangxi and Guangdong (Figure 4e). The northern part, which was primarily occupied by sporadic hot spots, was surrounded by several oscillating ones, while the latter laid in the eastern and southern parts. For rebellions in sporadic hot spots, there were more time steps during the second wet period, but oscillating hot spots included more battles during the last wet stage. Thus, the hot spots in the north and south presented different patterns. The hot spots during all dry phases were principally concentrated in the lower reaches of the Yangtze River–Yangtze River Delta and expanded southwestward to eastern Guizhou, part of Hunan and Guangxi, along with those from Guanzhong to the Central Plain (Figure 4f). Except for several diminishing hot spots in the north with surrounding sporadic ones, oscillating hot spots dominated the pattern. The difference between sporadic and oscillating hot spots may be attributed to the proportions of time steps in different dry stages. For diminishing hot spots, rebellions were less frequent in later periods, which implies the southward movement of rebellion focus through time.

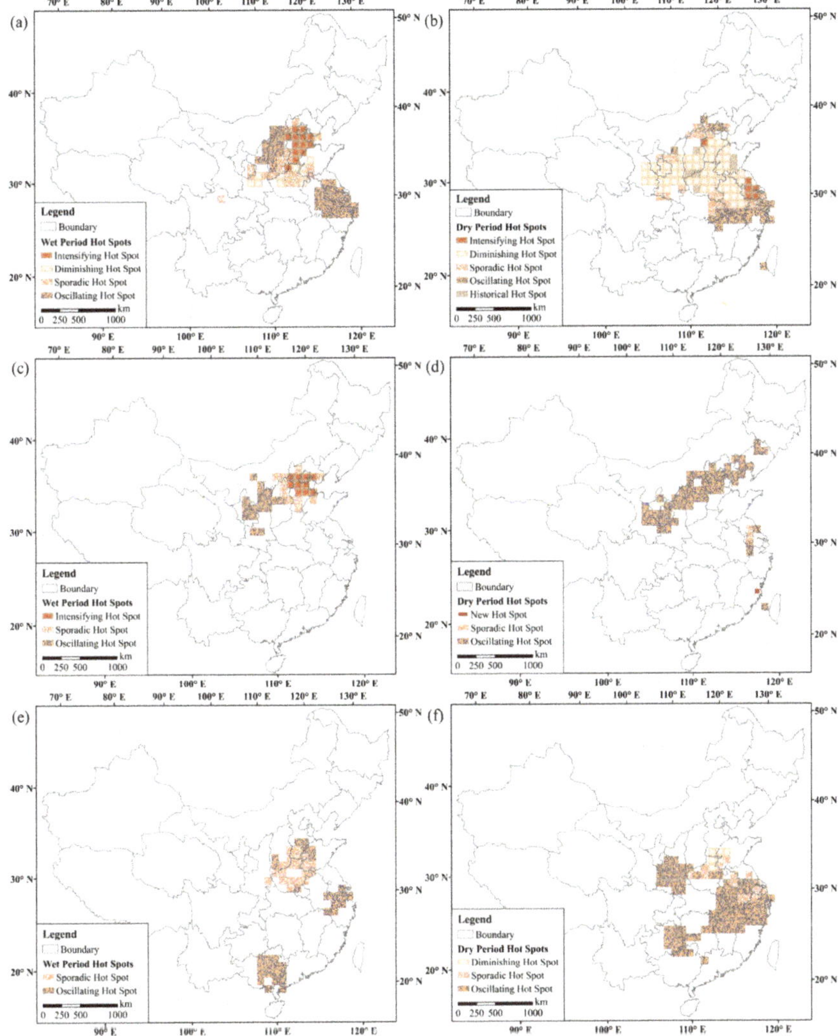

Figure 4. EHSA patterns of all wet and dry stages for three kinds of war. (**a**,**b**) All war, (**c**,**d**) agri-nomadic conflict, and (**e**,**f**) rebellion.

4. Discussion

4.1. Effect of Climate Change on the Hot Spot of War

In this study, we attempted to quantify the cluster effect of war and determine the spatiotemporal pattern of war hot spots in historical China by emphasizing a possible climatological root. On the whole, warm and wet (cold and dry) climate and the accompanying shifts in climatic especially temperature zones, may have brought about the north/west/northwestward (south/east/southeastward) movements of agricultural and pastoral zones. The spatial distributions of war hot spots, particularly the intensifying ones, varied correspondingly. Notably, different from traditional hot spots, those in this research contain temporal information (i.e., time steps/series in each bin). The proportions of time steps of later periods were larger in southern hot spots, which was consistent with the overall decline in temperature throughout the last two millennia (in particular, during the Little Ice Age (LIA), Figure 1a). In comparison, precipitation fluctuated as cycles without a significant trend (Figure 1b), which shows that the N–S disparities of hot spots may have been more temperature-dependent. The cyclic and long-term temperature variations and the resultant patterns of war hot spots in this work are similar to the findings in Zhang et al. [21], who used the Standard Deviational Ellipse as another spatial analytical tool but for the detection of directional characteristics/differences of war during three secular temperature cycles.

Specifically, as shown by the results in Figures 3 and 4, two categories of war—i.e., agri-nomadic conflict and rebellion, which responded to climate change differently, had different hot spot patterns. First, for agri-nomadic conflict, the strength of agricultural empire rose under warm and wet conditions, so that the central government could either afford expeditions toward Northwest China—i.e., the hinterland of steppes—and maintain control over the nomads there, or at least keep an equilibrium of military power against nomadic tribes/regimes along the borderline. This is why nearly all hot spots of agri-nomadic conflicts were concentrated in the north during such climatic phases. Conversely, cold and dry climate severely affected the economic bases of agricultural empires and nomadic regimes/tribes by reducing grain yields and herding resources, respectively. The weakened power of the central government meant that the military initiative held in warm and/or wet periods gave way to passive frontier defense against nomads [40]. Besides, driven by the cooling- and/or arid-induced environmental degradation, nomads had no choice but to migrate southward, and their conflicts with agriculturalists became increasingly intensive, even breaking through the boundaries and massively invading southward [41,42]. Noticeably, after comparing the patterns between Figures 3d and 4d, we observed many more hot spots in Southeast China during cold intervals than in dry periods, which implied that the effect of temperature on war hot spots may have been stronger than that of precipitation.

Second, with respect to rebellion, similar to the case of agri-nomadic conflict, warm and wet climate typically boosted the economy of agricultural empire, so that the government was able to initiate military campaigns toward the frontier. Moreover, the economic prosperity possibly resulted in rapid population growth, and the demands for expanding living space and relieving the ever-increasing population pressure would have needed to be met. As a result, accompanied by the outward conquest, mass migrations of agriculturalists also occurred during warm and wet intervals, which may have exacerbated the competitions for resources (especially for land) between the Han Chinese and ethnic minorities. Therefore, the emergence and spread of revolts with separate clusters in remote areas, such as the northwest of the Sichuan Basin, Taiwan, Guangdong, and Guangxi, could be interpreted. In comparison, the rebellion hot spots during cold and dry stages shifted inward and widely covered the middle–lower reaches of the Yangtze River, Yangtze River Delta, and the border among Guizhou, Hunan, and Guangxi, but still manifested in the Central Plain and part of the North China Plain, even extending westward to eastern Gansu, as the agricultural base of the Chinese Empire was heavily undermined and the frontier control over ethnic minorities loosened. Nonetheless, the Central Plain remained the core warring zone regardless of climatic phase, yet rebellions somewhat intensified during cold periods.

4.2. Contributions of Other Possible Factors

Our investigation into the linkage between climate and war hot spots does not exclude the possible contributions of other non-climatic factors that mediate the relationship. However, these factors are extremely difficult to spatially quantify from historical documents, in which they are often described ambiguously and fragmented in earlier ages. Yet, such limitations do not interfere with our discussion. Clearly, the hot spots (particularly the intensifying ones) of war overlapped with the developed and populous areas in China, e.g., the Guanzhong Basin, the Central–North China Plain, the Jianghuai region, and the Yangtze River Delta. Hence, this spatial consistency could be interpreted as population-pressure-led social contradiction, an important factor that was observed before [43–47]. Northern China, where the political center of the Chinese Empire was usually located, greatly benefited from warm and/or wet conditions. The rapid economic development and population growth in such an ecologically fragile agricultural–pastoral transition zone sooner or later deteriorated the environment and caused problems like soil erosion, desertification, and salinization, thereby lowering the land-carrying capacity and triggering resource struggles, and even armed conflict, within the Han Chinese peasants/nomads (e.g., during the Sixteen Kingdoms), or between agriculturalists and pastoralists. Likewise, as northern migrants fled nomadic invasions during cold and/or dry intervals, population increased and the economy flourished in southern China, followed by increasing numbers of conflicts and the emergence of the hot spot zone. Nevertheless, the spatial relationship between war and population/migration in ancient China needs further surveys.

As the agricultural basis in the north was damaged by temperature drops (C5–C7, i.e., the LIA) together with long-term drought (i.e., D3), migration also pointed to the southwest during the late imperial epoch. After the migration from Jiangxi to Huguang in the Ming dynasty, the more massive migration in the Qing dynasty (also known as "Huguang fills Sichuan") boosted the population of the entire southwest, since migrants also flooded into the Yunnan–Guizhou Plateau [48]. Overpopulation on these ecologically fragile mountainous karst landforms exhausted the land-carrying capacity and aggravated the tension between the Han and minorities. Consequently, the hot spots of war were concentrated in these regions. Apart from the population stress resulted from mass migration (sometimes launched by the government), governmental policy may have acted as a catalyst for conflict. For instance, throughout the Ming and Qing dynasties, the management and exploitation of southwestern China deepened. One momentous measure, the bureaucratization of native officers ("Gaitu Guiliu"), aiming at abolishing the hereditary local chieftain system (or "Tusi System") and directly assigning officials by the central government, was implemented. This policy worsened the relationships between the central government and ethnic minorities and ignited numerous revolts in that region. Hence, for rebellion (some were waged by the Miao, Dong, Zhuang, and Yao people), the concentrations of hot spots in the west of the Sichuan Basin, Guangxi, and the Guizhou Plateau could be ascribed to the strengthened rule of the government.

Another possible factor is pertinent to the geopolitical situation. The N–S regime confrontation, which was qualitatively elucidated in previous studies [22,23,49–52], is one of the most pronounced characteristics in Chinese history. It included two aspects: united empires that occupied most of the land area to the south of the Great Wall (China Proper) versus nomadic tribes/regimes that stayed in the steppes of Inner Asia during warm and wet phases, as well as two or more Han Chinese or Han versus nomadic regimes within China Proper during disintegrated eras and also under cold and dry conditions. Looking at Table 1, Table S2, Figure 3b, and Figure 4b, the hot spots (especially the intensifying ones) in the southeast can be partly explained by the latter aspect. Under the circumstances, battles were dominantly clustered in the vicinity of regime borders, or distributed along natural boundaries, such as the Yellow River (e.g., Northern Wei versus Liu Song during the Southern and Northern dynasties), the Qinling Mountains–Huai River (e.g., Wei versus Shu and Wu in the Three Kingdoms Period and Jin versus Southern Song), or farther south, the Yangtze River (e.g., Sui versus Chen).

5. Conclusions

This study is the first attempt to probe into long-term climate change (warm/wet versus cold/dry) and its association with the hot spots of different types of war (all war, agri-nomadic conflict, and rebellion) in imperial China. Looking at the connection from a spatiotemporal perspective with the aid of a few quantitative and visualization techniques, we conclude the following:

1. The cluster effect of war in ancient China was quantified by using the Global Moran's I, i.e., cells with large battle numbers tended to concentrate adjacently.
2. The Mann–Kendall trend test showed that, at any climatic mode, agri-nomadic conflict and rebellion basically increased over time, whereas the results for all war were almost insignificant.

3. Regarding EHSA, the hot spots for all war shifted northward and westward during warm and wet intervals, but toward the southeast in cold and dry conditions. For agri-nomadic conflict, hot spots were distributed along the boundary between agricultural and pastoral regimes in warm and wet phases, but reached as far as South China (due to nomadic invasions) during cold and dry stages. For rebellion, with the vicissitude of the Chinese Empire, hot spots spread outward in three to four groups in warm and wet climate but contracted inward during cold and dry periods.

4. EHSA satisfactorily reflected the pattern of war hot spots, on which temperature may have exerted more effect than precipitation.

Supplementary Materials: The following are available online at http://www.mdpi.com/2073-4433/11/4/378/s1, Figure S1: Comparisons of battle number counted within the grids of (a) 50 km, (b) 100 km, and (c) 200 km for each side of a cell, Figure S2: Geographical features of China mentioned in the main text, Table S1: History of China in the study period, Table S2a: N–S confrontation within China Proper during warm and cold phase, 5–1911 CE, Table S2b: N–S confrontation within China Proper during wet and dry phase, 1–1911 CE, Table S3: The Global Moran's *I* for different categories of war and climatic stages based on the grids of 50, 100, and 200 km.

Author Contributions: Conceptualization, D.D.Z.; Data curation, S.Z. and D.D.Z.; Formal analysis, S.Z.; Funding acquisition, D.D.Z.; Investigation, S.Z.; Methodology, S.Z. and D.D.Z.; Project administration, D.D.Z.; Resources, D.D.Z.; Supervision, D.D.Z. and J.L.; Validation, S.Z., D.D.Z., and J.L.; Visualization, S.Z.; Writing—original draft, S.Z.; Writing—review and editing, S.Z., D.D.Z., and J.L. All authors have read and agreed to the published version of the manuscript.

Funding: This research was funded by the Fund for Key Platform Construction Project–Special Project of High Level University Construction at Guangzhou University, grant number 290020363.

Acknowledgments: The authors thank the anonymous reviewers for their valuable comments on this manuscript.

Conflicts of Interest: The authors declare no conflict of interest.

References

1. Zhang, D.; Jim, C.; Lin, C.; He, Y.; Lee, F. Climate change, social unrest and dynastic transition in ancient China. *Chin. Sci. Bull.* **2005**, *50*, 137–144. [CrossRef]
2. Zhang, D.D.; Jim, C.Y.; Lin, G.C.; He, Y.; Wang, J.J.; Lee, H.F. Climatic change, wars and dynastic cycles in China over the last millennium. *Clim. Chang.* **2006**, *76*, 459–477. [CrossRef]
3. Zhang, D.D.; Zhang, J.; Lee, H.F.; He, Y. Climate change and war frequency in eastern China over the last millennium. *Hum. Ecol.* **2007**, *35*, 403–414. [CrossRef]
4. Lee, H.F.; Zhang, D.D. Changes in climate and secular population cycles in China, 1000 CE to 1911. *Clim. Res.* **2010**, *42*, 235–246. [CrossRef]
5. Zhang, D.D.; Lee, H.F. Climate change, food shortage and war: A quantitative case study in China during 1500–1800. *Catrina* **2010**, *5*, 63–71.
6. Zhang, Z.; Tian, H.; Cazelles, B.; Kausrud, K.L.; Bräuning, A.; Guo, F.; Stenseth, N.C. Periodic climate cooling enhanced natural disasters and wars in China during AD 10–1900. *Proc. R. Soc. B* **2010**, *277*, 3745–3753. [CrossRef]
7. Bai, Y.; Kung, J.K.S. Climate shocks and Sino-nomadic conflict. *Rev. Econ. Stat.* **2011**, *93*, 970–981. [CrossRef]
8. Lee, H.F.; Zhang, D.D. A tale of two population crises in recent Chinese history. *Clim. Chang.* **2013**, *116*, 285–308. [CrossRef]
9. Lee, H.F. Climate-induced agricultural shrinkage and overpopulation in late imperial China. *Clim. Res.* **2014**, *59*, 229–242. [CrossRef]
10. Pei, Q.; Zhang, D.D. Long-term relationship between climate change and nomadic migration in historical China. *Ecol. Soc.* **2014**, *19*, 68. [CrossRef]
11. Zheng, J.; Xiao, L.; Fang, X.; Hao, Z.; Ge, Q.; Li, B. How climate change impacted the collapse of the Ming dynasty. *Clim. Chang.* **2014**, *127*, 169–182. [CrossRef]
12. Fang, X.; Su, Y.; Yin, J.; Teng, J. Transmission of climate change impacts from temperature change to grain harvests, famines and peasant uprisings in the historical China. *Sci. China Earth Sci.* **2015**, *58*, 1427–1439. [CrossRef]

13. Xiao, L.; Fang, X.; Zheng, J.; Zhao, W. Famine, migration and war: Comparison of climate change impacts and social responses in North China between the late Ming and late Qing dynasties. *Holocene* **2015**, *25*, 900–910. [CrossRef]

14. Zhang, D.D.; Pei, Q. Gone with winds: A quantitative analysis of battlefield locations and climate shifts in imperial China. In *Geo-Strategy and War: Enduring Lessons for Australian Army*; Dennis, P., Ed.; Big Sky Publishing: Newport, NSW, Australia, 2015; pp. 193–211.

15. Zhang, D.D.; Pei, Q.; Lee, H.F.; Zhang, J.; Chang, C.Q.; Li, B.; Li, J.; Zhang, X. The pulse of imperial China: A quantitative analysis of long-term geopolitical and climatic cycles. *Glob. Ecol. Biogeogr.* **2015**, *24*, 87–96. [CrossRef]

16. Lee, H.F.; Zhang, D.D.; Pei, Q.; Jia, X.; Yue, R.P. Demographic impact of climate change on northwestern China in the late imperial era. *Quat. Int.* **2016**, *425*, 237–247. [CrossRef]

17. Liu, L.; Su, Y.; Fang, X. Wars between farming and nomadic groups from Western Han dynasty to Qing dynasty in north China and relationship with temperature change. *J. Beijing Normal Univ. (Nat. Sci.)* **2016**, *52*, 450–457, (In Chinese with English abstract).

18. Lee, H.F.; Zhang, D.D.; Pei, Q.; Jia, X.; Yue, R.P. Quantitative analysis of the impact of droughts and floods on internal wars in China over the last 500 years. *Sci. China Earth Sci.* **2017**, *60*, 2078–2088. [CrossRef]

19. Pei, Q.; Lee, H.F.; Zhang, D.D. Long-term association between climate change and agriculturalists' migration in historical China. *Holocene* **2017**, *28*, 208–216. [CrossRef]

20. Zhang, S.; Zhang, D.D. Population-influenced spatiotemporal pattern of natural disaster and social crisis in China, AD1–1910. *Sci. China Earth Sci.* **2019**, *62*, 1138–1150. [CrossRef]

21. Zhang, S.; Zhang, D.D.; Li, J.; Pei, Q. Secular temperature variations and the spatial disparities of war in historical China. *Clim. Chang.* **2020**. [CrossRef]

22. Song, J. The geographical pivot of war in ancient China. *J. Cap. Normal Univ. (Soc. Sci. Ed.)* **1994**, *4*, 1–10. (In Chinese)

23. Rao, S. *Layout of the World: General Situation of Military Geography in Ancient China*; People's Liberation Army Press: Beijing, China, 2002. (In Chinese)

24. Wang, X. Spatial distribution and focus area transfer of war and battle in poems of the early-prosperous Tang dynasty. *Data Cult. Educ.* **2009**, *18*, 16–18. (In Chinese)

25. Leng, H. Research on the Transfer of the Northern Border Trouble in Ancient China from Qin and Han to Ming Dynasties. Master's Thesis, Liaoning University, Shenyang, China, May 2011.

26. Zhu, Y.; Newsam, S. Spatio-temporal sentiment hotspot detection using geotagged photos. In Proceedings of the 24th ACM SIGSPATIAL International Conference on Advances in Geographic Information Systems, Burlingame, CA, USA, 31 October–3 November 2016; ACM: New York, NY, USA, 2016. [CrossRef]

27. Sadler, R.C.; Pizarro, J.; Turchan, B.; Gasteyer, S.P.; McGarrell, E.F. Exploring the spatial-temporal relationships between a community greening program and neighborhood rates of crime. *Appl. Geogr.* **2017**, *83*, 13–26. [CrossRef]

28. Coyan, J.A.; Zientek, M.L.; Mihalasky, M.J. Spatiotemporal analysis of changes in lode mining claims around the McDermitt Caldera, northern Nevada and southern Oregon. *Nat. Resour. Res.* **2017**, *26*, 319–337. [CrossRef]

29. Hosseini, S.M.; Parvin, M.; Bahrami, M.; Karami, M.; Olfatifar, M. Pattern mining analysis of pulmonary TB cases in Hamadan Province: Using space-time cube. *Int. J. Epidemiol. Res.* **2017**, *4*, 111–117.

30. Bunting, R.J.; Chang, O.Y.; Cowen, C.; Hankins, R.; Langston, S.; Warner, A.; Yang, X.; Louderback, E.R.; Sen Roy, S. Spatial patterns of larceny and aggravated assault in Miami-Dade County, 2007–2015. *Prof. Geogr.* **2017**, *70*, 34–46. [CrossRef]

31. Harris, N.L.; Goldman, E.; Gabris, C.; Nordling, J.; Minnemeyer, S.; Ansari, S.; Lippmann, M.; Bennett, L.; Raad, M.; Hansen, M.; et al. Using spatial statistics to identify emerging hot spots of forest loss. *Environ. Res. Lett.* **2017**, *12*, 024012. [CrossRef]

32. Ge, Q.; Hao, Z.; Zheng, J.; Shao, X. Temperature changes over the past 2000 yr in China and comparison with the Northern Hemisphere. *Clim. Past.* **2013**, *9*, 1153–1160. [CrossRef]

33. Mann, M.E.; Zhang, Z.; Hughes, M.K.; Bradley, R.S.; Miller, S.K.; Rutherford, S.; Ni, F. Proxy-based reconstructions of hemispheric and global surface temperature variations over the past two millennia. *Proc. Natl. Acad. Sci. USA* **2008**, *105*, 13252–13257. [CrossRef]

34. Editorial Committee of Chinese Military History. *Tabulation of Wars in Ancient China*; People's Liberation Army Press: Beijing, China, 1985. (In Chinese)
35. Tan, Q. *Historical Atlas of China*; SinoMaps Press: Beijing, China, 1982. (In Chinese)
36. How Spatial Autocorrelation (Global Moran's *I*) Works. Available online: https://desktop.arcgis.com/en/arcmap/latest/tools/spatial-statistics-toolbox/h-how-spatial-autocorrelation-moran-s-i-spatial-st.htm (accessed on 26 August 2019).
37. Mann, H.B. Nonparametric tests against trend. *Econometrica* **1945**, *13*, 245–259. [CrossRef]
38. Kendall, M.G.; Gibbons, J.D. *Rank Correlation Methods*, 5th ed.; Griffin: London, UK, 1990.
39. How Hot Spot Analysis (Getis-Ord *Gi**) Works. Available online: https://desktop.arcgis.com/en/arcmap/latest/tools/spatial-statistics-toolbox/h-how-hot-spot-analysis-getis-ord-gi-spatial-stati.htm (accessed on 26 August 2019).
40. Wang, X.; Chen, F.; Zhang, J.; Yang, Y.; Li, J.; Hasi, E.; Zhang, C.; Xia, D. Climate, desertification and the rise and collapse of China's historical dynasties. *Hum. Ecol.* **2010**, *38*, 157–172. [CrossRef]
41. Fang, J. The impacts of climatic change on the Chinese migrations in historical times. *Geogr. Environ. Res.* **1989**, *1*, 39–46, (In Chinese with English abstract).
42. Wang, H. The relationship between the migrating south of the nomadic nationalities in north China and the climatic changes. *Sci. Geogr. Sin.* **1996**, *16*, 274–279, (In Chinese with English abstract).
43. Ho, P.T. *Studies on the Population of China, 1368–1953*; Harvard University Press: Cambridge, MA, USA, 1959.
44. Webster, D. Warfare and evolution of state—reconsideration. *Am. Antiq.* **1975**, *40*, 464–470. [CrossRef]
45. Zhao, W.; Xie, S. *History of Chinese Population*; People's Press: Beijing, China, 1988. (In Chinese)
46. Carneiro, R.L. The transition from quantity to quality: A neglected causal mechanism in accounting for social evolution. *Proc. Natl. Acad. Sci. USA* **2000**, *97*, 12926–12931. [CrossRef]
47. Lee, H.F.; Fok, L.; Zhang, D.D. Climatic change and Chinese population growth dynamics over the last millennium. *Clim. Chang.* **2008**, *88*, 131–156. [CrossRef]
48. Yang, Y. Migration and frontier society: Regulation of the Han-minority contradiction in Yunnan–Guizhou region during the Qing dynasty under the perspective of social control. *J. Yunnan Normal Univ. (Humanit. Soc. Sci.)* **2017**, *49*, 9–19. (In Chinese)
49. Shi, N. Discussion about the situations of west–east and north–south confrontations in historical China. *J. Chin. Hist. Geogr.* **1992**, *1*, 57–112. (In Chinese)
50. Hu, A. *Place with Strategic Importance: An Outline of Military Geography in Historical China*; Hehai University Press: Nanjing, China, 1996. (In Chinese)
51. Yao, X. A brief discussion on the evolution of the spatial orientation of war in ancient China. *Mil. Hist.* **2007**, *4*, 58–60. (In Chinese)
52. Yao, X. Study on the geopolitical model of wars in ancient Chinese dynasties. *Hum. Geogr.* **2007**, *1*, 125–128, (In Chinese with English abstract).

Article

Regional Interactions in Social Responses to Extreme Climate Events: A Case Study of the North China Famine of 1876–1879

Xianshuai Zhai [1,2], Xiuqi Fang [1,2] and Yun Su [1,2,*]

[1] Faculty of Geographical Science, Beijing Normal University, Beijing 100875, China; 201831051021@mail.bnu.edu.cn (X.Z.); xfang@bnu.edu.cn (X.F.)

[2] Key Laboratory of Environmental Change and Natural Disaster, Ministry of Education, Beijing Normal University, Beijing 100875, China

* Correspondence: suyun@bnu.edu.cn

Received: 14 March 2020; Accepted: 8 April 2020; Published: 16 April 2020

Abstract: The North China Famine of 1876–1879, known in Chinese as the Dingwu qihuang (丁戊奇荒), is a famous case of drought-induced famine in Chinese history. The purpose of this paper is to provide empirical and historical evidence for understanding the impacts of extreme climate events and major disasters and the mechanisms of adaptation. From the aspects of famine-related migration and the allocation of relief money and grain, the regional interactions in social responses to extreme climate events were analyzed. This paper collected 186 records from historical documents. Regarding the regions as the nodes and the relationships between regions as the links, the spatial patterns of famine-related migration and the allocation of money and grain from 1877 to 1878 were rebuilt. The results show that, firstly, famine-related migration appeared to be spontaneous and short-distanced, with the flow mainly spreading to the surrounding areas and towns. Secondly, as a state administrative action, the relief money and grain from the non-disaster areas were distributed to the disaster areas. However, the distribution of relief grain affected the equilibrium of the food market in non-disaster areas, which led to fluctuations in food prices.

Keywords: drought; regional interaction; North China Famine of 1876–1879

---

1. Introduction

Scholars have been studying the complex interactions between climate change and human history [1], and history is key to understanding the present and future. One of the major research themes of the Past Global Changes (PAGES) project focuses on the social impacts of historic extreme climate events and the responses, as well as the mechanisms and processes of past human-climate-ecosystem interactions at multiple spatial and temporal scales. It aims to enhance our understanding of the influence of contemporary climate change and the adaptation of human society. This is an international effort to coordinate and promote past global change research [2].

Among the studies on the impacts of and responses to the historic extreme climate events, most cases discussed the relationship between the human social system and the climate-environment system in the same region, but rarely involve regional correlations or common responses. In fact, when the impact of extreme climate events exceeds the regional carrying capacity, not only the affected areas, but also the initially non-affected areas can be influenced. There will be a common response in both affected areas and non-affected areas. For example, from 1813 to 1815, floods and droughts struck many countries of Europe, resulting in crop failures. Approximately 8000 refugees from Southern Germany migrated to Russia in the east. France, Italy, and the Netherlands imported food grains from Egypt, Russia, the United States, and some other regions [3]. In Australia, during the period of 1800–1945, in the face of drought or floods, the social responses included the relocation of towns and the establishment of dams to coordinate water consumption in the upstream and downstream areas of the river basin [4]. In the southern part of North America, there were several severe drought events in the 9th to 14th centuries, and people there abandoned the infrastructures and migrated [5]. In China, from 1560 to 1890, it was at the height of the Little Ice Age that the climate fluctuated violently. The social responses, including famine-related migration and the allocation of money and grain, can be observed [6–12].

The presented case studies show that regional interaction has become an essential way of social response to historic extreme climate events. However, there is a lack of research on the characteristics, processes, and mechanisms of these regional interactions. Given the impact of extreme climate events and the social response could generate a complex multidirectional network in time and space [13], it is necessary to contribute to the research on the social response mechanism from the perspective of regional interaction.

In the present, global connectivity is continuously enhancing, and regional integration is deepening. It can be inferred that at the local, national, regional and global levels, the possibility of being affected by extreme events is increasing. According to the report of the American Meteorological Association, climate change is closely related to extreme events, and these events will seriously threaten the social economy and human life [14]. Therefore, inter-regional coordinated responses are urgently needed. IPCC's report pointed out that risk transfer and sharing will be an effective way of social response [15]; yet considering the interdependencies between regional economic and social systems, it may have opposite effects on different regions, which means the disaster risks could be either reduced or even amplified for a certain region involved [16,17]. Although it is impossible to reproduce the exact results of the response to the past events, the mechanisms, experience, and lessons of regional interactions in response to the historic extreme climate events are still be of an essential reference value.

Using documentary evidence to study past extreme climate events has become a recognized method [18], which emphasizes China's advantages in researching the social impacts of and response mechanisms to past climate change. On the one hand, the monsoon climate in China is characterized by its instability, and the traditional agriculture-based economy made the socio-economic system significantly sensitive to the changes in climate. On the other hand, China owns abundant and continuous documentary records left by its long history, such as historical books, local chronicles, archived documents, private diaries, etc. Besides, newspapers, which were first published in China in the early 19th century, are the documentary records with a high temporal resolution. They can not only be used to reconstruct the precipitation, temperature and other weather conditions in history [19,20], but also to explore the whole development process of historic extreme climate events within the socio-economic systems [18,21].

In this paper, focusing on the famine-related migration and the allocation of money and grain, the spatial and temporal features of the regional interactions in response to the North China Famine of 1876–1879 (known in Chinese as the "Dingwu qihuang" or the "Incredible Famine of 1877–1878") are analyzed. Combining with exploration on the after-effects of the famine on both disaster areas and non-disaster areas, the results provide the empirical evidence for understanding social response mechanisms from the perspective of inter-regional linkages.

2. Data Sources and Research Methods

2.1. Case Selection

North China is a region with a temperate monsoon climate, prone to drought in spring, summer and autumn. In the past 2000 years, there have been 227 extreme drought events in North China. Droughts occurred more frequently in 150–200 A.D., 550–800 A.D., 1050–1100 A.D., and 1850–1900 A.D. [22].

From 1876 to 1879, five provinces in North China, namely, Shanxi, Henan, Shaanxi, Hebei, and Shandong (Figure 1), suffered a severe drought. The reconstructed precipitation (wet/dry) series indicated that it was the most severe drought in this region in the past 300 years [23]. Sea surface temperature anomalies in the eastern Pacific region and intense El Niño events [24] had resulted in the weakening of the East Asian monsoon and precipitation variability, which were the direct causes of this drought event. It had global effects, with several regions experiencing extreme drought at the same time, including Australia, Europe, North America and South America [25–28].

Figure 1. Study area in the paper. In the North China Plain: Beijing, Hebei, Shandong, and Henan; In the Loess Plateau: Shanxi, Shaanxi, and Gansu; In the Yangtze River Basin: Sichuan, Guizhou, Hubei, Hunan, Anhui, Jiangxi; Jiangsu, and Zhejiang; andIn the Jiangnan region: Fujian, Guangxi and Guangdong.

In China, the year 1877 was classified as a "Ding" year, and 1878 was classified as a "Wu" year. Because the worst period was in 1877–78, this extreme drought event was historically known as the "Dingwu qihuang". In 1877, 20% of the villages in Shanxi Province experienced harvest failures, and in the central Shaanxi Province, the harvest rate of grains during the fall harvest season was merely 30%. The famine reached its peak in 1877, with Shanxi and Henan worst affected [29]. Worse still, an epidemic occurred soon after and had spread over a large area during the spring and summer of 1878.

Regarding the social factors, frequent warfare in the late Qing Dynasty, fiscal crisis, and overburdened tenants aggravated the severity of disaster [30–32], leading to severe damage to productivity, homelessness, and social crisis [23].

In the end, approximately 160 to 200 million people were affected by the drought, and about 9.5 to 13 million people died from famine and disease. Many worst-hit counties in Shanxi and Henan provinces had lost over 50% of their population, with the death toll passed 5 million and 1.8 million respectively [33].

2.2. Study Area

The study area was divided into two parts (Figure 1). One is the affected areas. They are located mainly in the Loess Plateau and the North China Plain, which are the major wheat-growing areas in China, with a long development history and large population, containing the above-mentioned five drought-stricken provinces. The capital city and the political center of China, Beijing, is also in this region. This region is the target of the relief efforts carried out by the Qing government. The Qing state's responses to the famine consisted of a variety of strategies, such as allocating relief silver and grain and reducing or canceling taxes.

The other part is the south region, containing the Yangtze River Basin and Jiangnan region. It is the place where the economic center of China in the Qing Dynasty was located, and the resources for the disaster relief mainly came from. The landforms of this region are featured with the plain area along the middle–lower reaches of the Yangtze River, and the hilly area in the southeastern part. Different from the north region, this area is with a subtropical monsoon climate. Good hydrothermal conditions and well-developed water systems are conducive to the growth of rice, wheat, and a variety of cash crops, as well as the development of forestry and fishing.

2.3. Data Sources

The data about the North China Famine of 1876–1879 were extracted from Qing Shi Lu [34], Shenbao [35], Disaster annals in recent China [36], and Qing Tong Jian [37] (Table 1).

Qing Shi Lu is a long-term compilation of the chronicles of the Qing dynasty. It contains a total of 4363 volumes. The materials in Qing Shi Lu are originally from the official documents of the Imperial Cabinet and other departments, the pieces of writing from the National Historical Archives, and some first-hand materials such as the emperor's anthology and handwriting [10]. The historical materials in Qing Shi Lu are of exceptionally high value.

By 1876, the Shenbao had established itself as a commercially successful newspaper that carried the only public and serious discussion of many public issues in China [38]. From 1876–79, Shenbao's critical coverage of the famine focused not only on the five hardest-hit northern provinces but also on some other areas influenced by the drought event [39].

Disaster annals in recent China systematically and chronologically expounded on the natural disasters in China from 1840 to 1919, combined with explicit analyses on the time, location, extent, causes and social influences of various natural disasters, as well as the effectiveness and gaps of disaster mitigation measures [40]. The data is of high quality and reliability.

In addition to the above sources, data source about Dingwu qihuang is also supplemented by Qing Tong Jian. Its collection of historical materials is complete and reliable, and it discusses in detail politics, society, finance, economy, transportation, war, etc.

Table 1. Information on sources of the data about the North China Famine of 1876–1879.

Data Sources	Temporal Coverage	Language	Access
Qing Shi Lu	1876–1879	Chinese	ISBN 978710105626
Shenbao	1876–1879	Chinese	https://www.neohytung.com/
Disaster annals in recent China	1876–1879	Chinese	ISBN 9787535510839
Qing Tong Jian	1876–1879	Chinese	ISBN 9787203039075
Food Price Database in the Qing Dynasty	1876–1879	Chinese	http://mhdb.mh.sinica.edu.tw/foodprice/
Zhang's research	1877–1878	English	doi:10.3724/SP.J.1248.2010.00091

After the removal of redundant records, a total of 186 historical records were extracted and classified (Table 2). We excerpted 96 records from Qing Shi Lu, 70 from Shenbao, 13 from Disaster annals in recent China, and 7 from Qing Tong Jian. Figure 2 shows that 1877–1878 is the key period, with significantly more records identified for the famine event.

Table 2. Classification of the historical records for the North China Famine of 1876–1879.

Contents	Number of Records	Records Indicated Directions	Records Indicated Quantities
Famine-related migration	56	30	6
Allocation of money and grain	74	74	34
Social unrest	56	56	7
Total	**186**	**160**	**47**

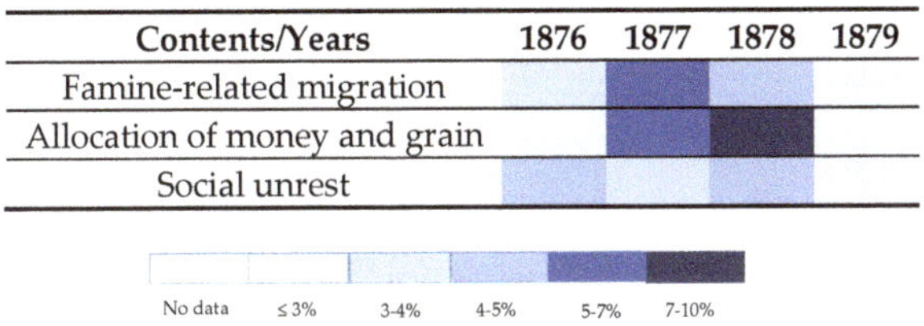

Contents/Years	1876	1877	1878	1879
Famine-related migration				
Allocation of money and grain				
Social unrest				

No data   ≤ 3%   3-4%   4-5%   5-7%   7-10%

Figure 2. Percentage of days with records in each year from 1876–1879.

This paper selected data of wheat prices from the Food Price Database in the Qing Dynasty [41] (Table 1). The database is based on historical documents in the First Historical Archives of China and the Institute of Economics of the Chinese Academy of Social Sciences.

The data on the famine-struck area and plague-infested area is from Zhang's research [42] (Table 1), which was based on climate records extracted from historical documents in "A compendium of Chinese meteorological records of the last 3000 Years (in Chinese)" [43].

2.4. Information Extraction

We extracted information on key activities, noting down the general processes, times and places of occurrence, and the numbers of starving migrants, or the amount of relief silver and grain. The times in the documents were converted into Gregorian calendar time format with monthly temporal resolution. The locations were recorded based on the current provincial-level and prefecture-level administrative divisions in China.

In the Qing Dynasty, grain was measured in piculs ("dan" in Chinese), and silver used in monetary transactions was measured in taels ("liang"). In the 19th century, 1 picul of rice weighed 60 kg [29]. 1 tael of silver was equal to $29.6 [44], converted according to the purchasing power of silver at that time.

Firstly, we identified the records of famine-related migration, for example, "(1878) this year, there was a severe drought in Henan, and starving people moved to Xuzhou in search of food [35]" (Figure 3a). Then the migration route was noted down: Henan→Xuzhou.

Secondly, the records of silver and grain allocation were also extracted. "(1877) when the severe drought struck Shanxi and Henan, the central government allocated 280,000 taels of silver to Shanxi and 120,000 taels of silver to Henan. Equally, 40,000 piculs of grain from the granaries in Anhui and Jiangsu shall be allocated to Shanxi [34]" (Figure 3b). The silver allocation was noted down: "Beijing→Shanxi, 280,000 taels; Beijing→Henan, 120,000 taels". The grain allocation: "Jiangsu/Anhui→Shanxi, 40,000 piculs". The grain in the paper refers to wheat and rice, the two major food staples in China.

Finally came to the records of social unrest during the famine. It includes revolt, banditry, and insurrection, e.g., "(1878) the bandits rebelled in Shanzhou [34]" (Figure 3c). In this record, the site of the unrest event was the Shanzhou district, which is now the Sanmenxia city of Henan Province.

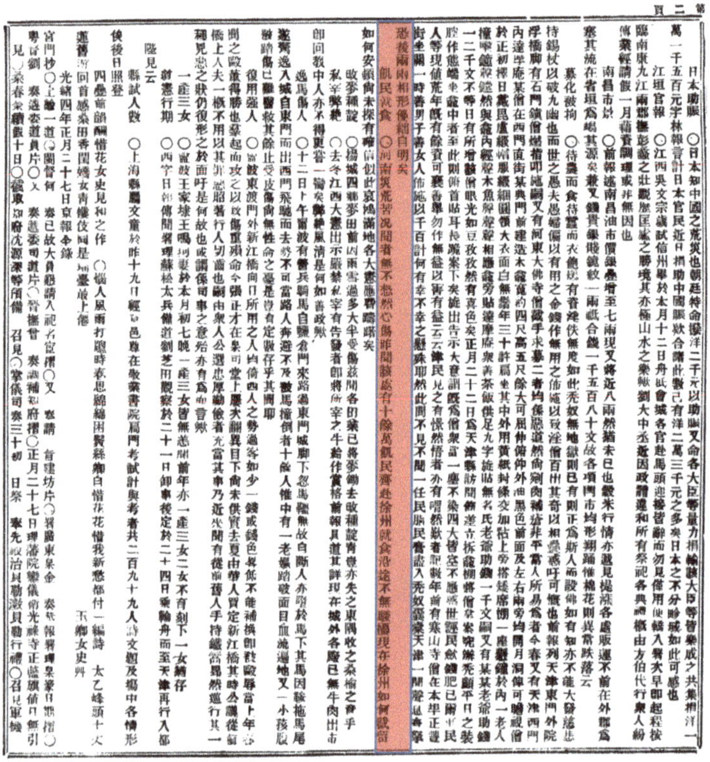

(**a**)

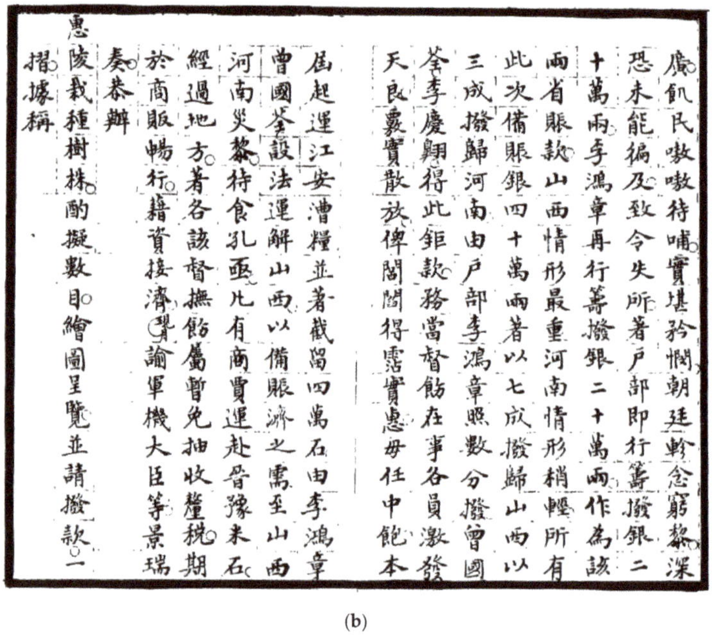

處飢民嗷嗷待哺實堪矜憫朝廷軫念窮黎深
恐未能徧及致令失所著戶部即行籌撥銀二
十萬兩李鴻章再行籌撥銀二十萬兩作為該
兩省賑款山西情形最重河南情形稍輕所有
此次備賑銀四十萬兩著以七成撥歸山西以
三成撥歸河南由戶部李鴻章照數分撥曾國
荃李慶翱得此鉅款務當督飭在事各員激發
天良覈實散放俾閭閻得霑實惠毋任中飽本
屆起運江安漕糧並著截留四萬石由李鴻章
曾國荃設法運解山西以備賑濟之需至山西
河南災黎待食孔亟凡有商賈運赴晉豫米石
經過地方著各該督撫飭屬暫免抽收釐稅期
於商販暢行藉資接濟將此諭軍機大臣等景瑞
奏恭辦
惠陵栽種樹株酌擬數目繪圖呈覽並請撥款一
摺據稱

(**b**)

Figure 3. *Cont.*

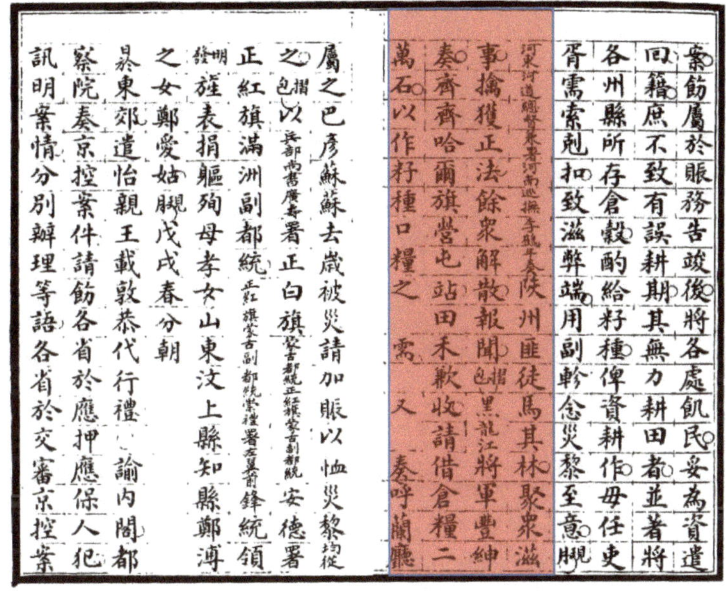

(**c**)

Figure 3. Historical records related to the North China Famine of 1876–1879 (**a**) is from Shenbao, about the famine-related migration; (**b**) is from Qing Shi Lu, about the allocation of money and grain; and (**c**) is also from Qing Shi Lu, about social unrest.

2.5. Spatial Analysis

To demonstrate the regional interactions in the social responses, the origins and destinations of the starving migrants were regarded as nodes, and the migration flows were regarded as the links. A spatial network of the famine-related migration in 1877–1878 was built (Figure 4). Similarly, taking the provinces where silver and grain transferred out and in as the nodes, the inter-provincial transfers as the links, and the number of transfers as the weight, a weighted network of money and grain allocation in 1877–1878 was built (Figure 5). In the networks, the degree refers to the number of direct links connected with each node. It can reflect the number or range of connections between different areas. Using ArcGIS, the lengths (distances) of migration and allocation flows can be calculated.

Besides, regarding the sites where the social unrest events took place as the core, point density analysis was conducted in ArcGIS. In the analysis, a particular sample point is set as the center, then a neighborhood is defined around each center with a certain value of search radius. After that, points of different weight values that fall within the neighborhood are identified. Points are weighted higher if they are closer to the centers. The weights gradually reduce until they drop to zero at the edge of the neighborhood. The density value is the sum of the values of the sample points divided by the area of the neighborhood [45,46]. In this paper, the unrest occurrence sites are the centers, and the search radius is 100 km. The unrest density value is divided into eight levels according to the number of unrest events per 10,000 square kilometers. The frequency of occurrence and the spillover effects of social unrest were analyzed.

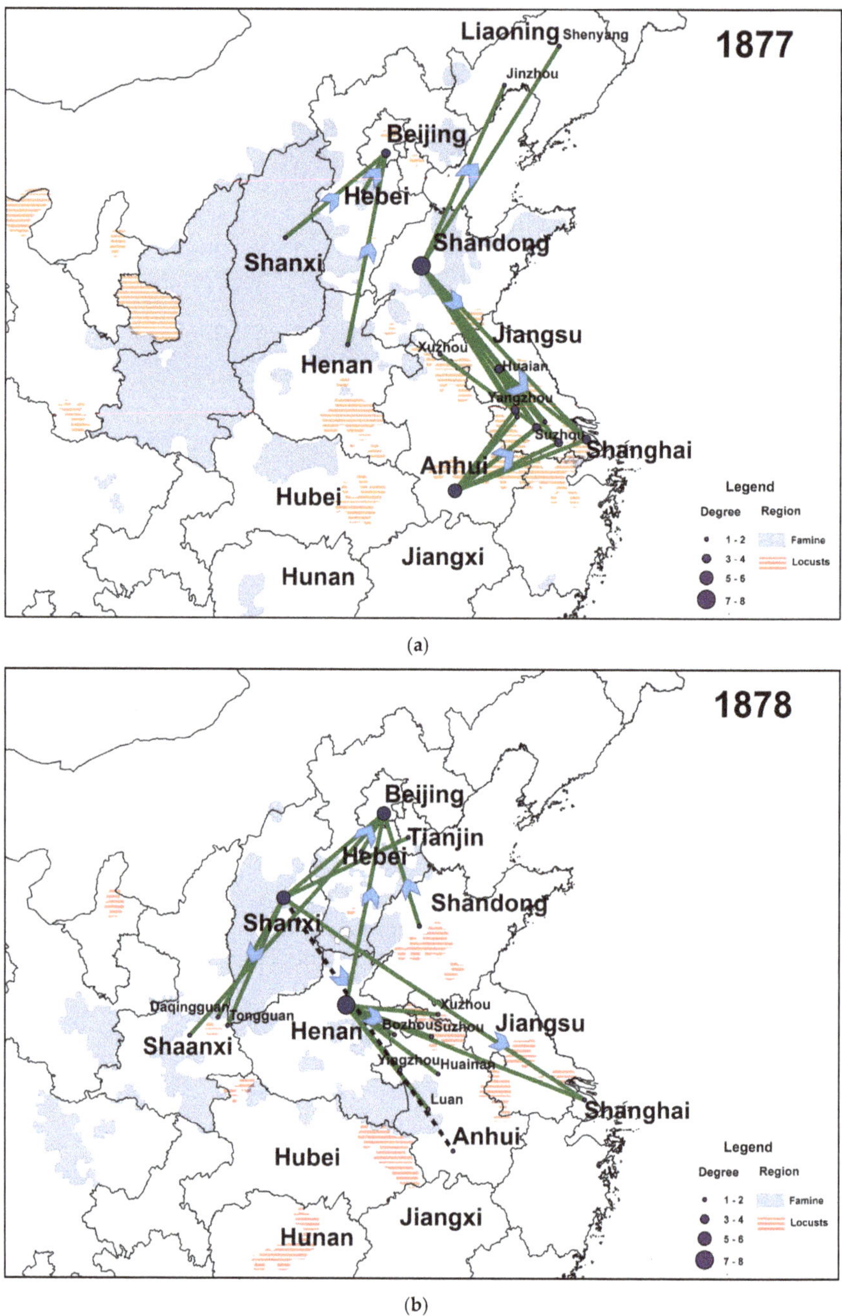

(a)

(b)

Figure 4. (**a**) Origins, destinations and sizes of the famine-related migration in 1877. (**b**) Origins, destinations and sizes of the famine-related migration in 1878. Straight line: migration flows with exact locations of destinations; Dotted line: migration flows with approximate locations of destinations; Degree: the number of direct links connected with a node.

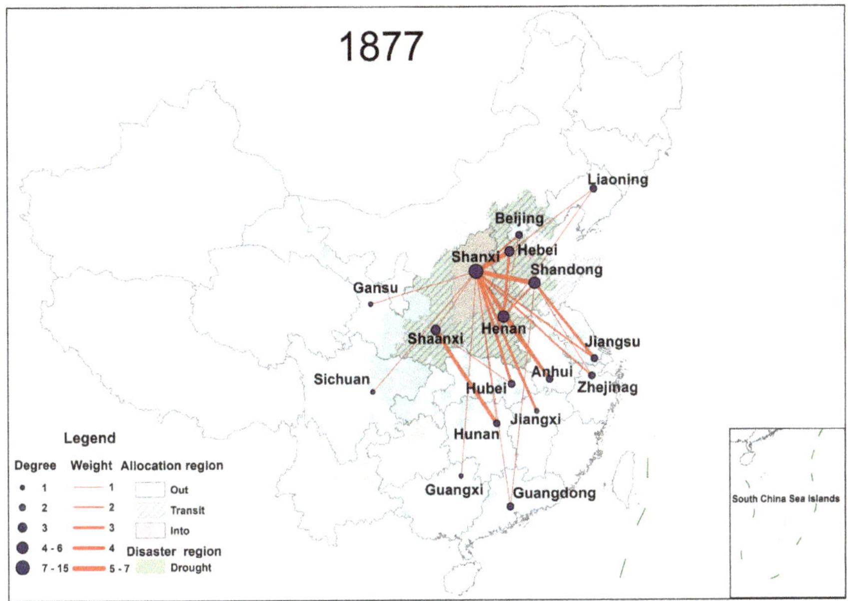

(**a**)

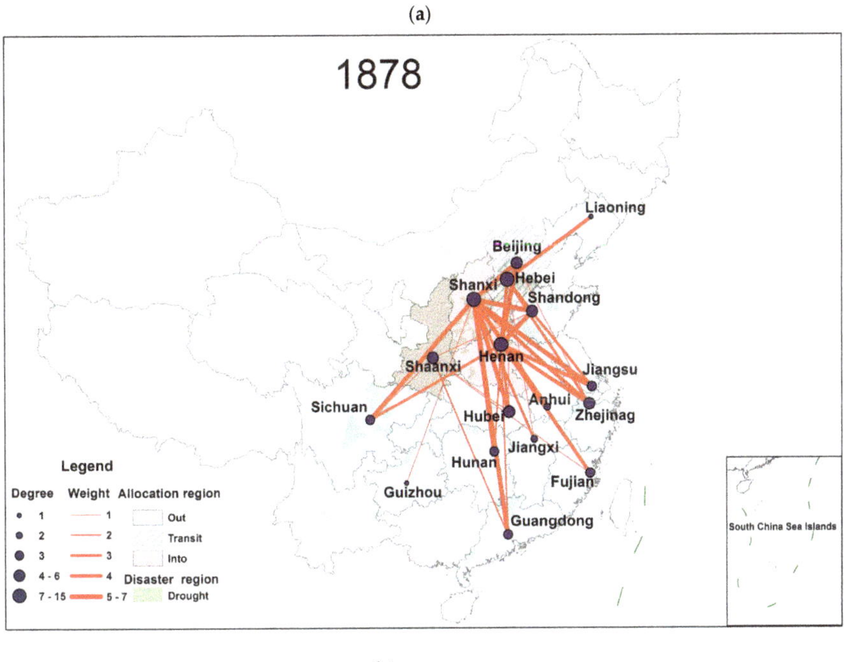

(**b**)

Figure 5. (**a**) Spatial characteristics of grain and monetary allocations in 1877. (**b**) Spatial characteristics of grain and monetary allocations in 1878. Degree: the number of direct links connected with a node; Weight: the number of inter-provincial transfers.

3. Results

3.1. Regional Interactive Response and Characteristics

3.1.1. Spatial Characteristics of Famine-Related Migrations

Historically, Chinese farmers have had a tied connection to their homeland and farmland. Large-scale population migration is temporary and spontaneous in the period of extreme climate events. The migration is often driven by famine, plague and other events triggered by major meteorological disasters [6]. The transition from being starving to homeless is the result of the local social system losing its ability to adapt and the failure of individual survival strategies.

Famine victims tended to migrate from the hardest-hit areas to the nearest slightly-impacted areas, and non-disaster areas (Figure 4). In 1877, migrants left from Shandong and Anhui to the south of Jiangsu. Some people left from Shandong to Liaoning, while others moved from Shanxi, Henan, and Hebei to Beijing. The victims migrated between 130 and 766 km (straight line distance, the same below), with an average migration distance of approximately 429 km. In 1878, famished people in Henan moved to Beijing and Anhui. Some people in Shandong, Hebei, and Shaanxi moved to Beijing, and others in Shanxi moved to Hebei, Beijing and Anhui, as well as to the Daqingguan and Tongguan in Shaanxi. Famine victims migrated between 130 and 1080 km, with an average migration distance of approximately 460 km.

In this historical period, limited by poor traffic conditions and the physical weakness, the spontaneous famine-related migration flows just spread from the disaster areas to the nearby areas. In 1877–1878, the harvest rate of grains in Anhui was only about 50%, while in Jiangxi, Hubei, and Hunan, which are farther away from the disaster areas, the harvest rate reached over 60% or 70% (Table 3). Although with relatively low harvest rate, Anhui was still one of the major destinations in the famine-related migration due to its proximity to the disaster areas. Meanwhile, the migration directions of the famished people appeared to be relatively fixed and stable, indicating the influences of regional politics, dissemination of the relief information, and customs and traditions [47].

Table 3. Harvest rate of grains in different provinces [48].

Year	Anhui		Jiangxi		Hubei		Hunan	
	Summer Harvest	Autumn Harvest	Summer Harvest	Autumn Harvest	Summer Harvest	Autumn Harvest	Summer Harvest	Autumn Harvest
1877	50%+	50%+	70%+	70%+	60%+	50%+	60%+	70%+
1878	50%+	50%+	60%+	60%+	60%+	60%+	60%+	70%+

There was a visible difference in the economic conditions between towns and villages in China during the Qing dynasties. Under the circumstance of the rural economic decline, cities and towns became the destinations of the migrants. Soup kitchens and shelters there offered more chances of survival. For example, the capital city at that time, Beijing, is close to the affected provinces, and it had become a popular destination for the starving migrants. Besides, some other places attracted migrants due to their advantageous geographical location and sophisticated traffic system. For instance, in 1878, the whole area of Shaanxi Province suffered a severe drought, but its two towns, Tongguan and Daqingguan, had experienced the entry of famished people from the Shanxi Province on the east. Lying close to the shared borders of Shaanxi, Shanxi and Henan provinces, these two towns were distribution centers of relief goods and had been of strategic importance since ancient times.

3.1.2. Spatial Characteristics of Money and Grain Allocations

In China, a state unified by centralized political power, the allocation of money and grain is a state administrative action. It is normal that the central government provides emergency financial assistance and food aid to disaster areas. The relief silver and grain are mainly from state banks and granaries, which are supplied by every locality on a regular basis. Besides, some resources in the non-disaster areas could be requisitioned as emergency relief by the central government for the duration of the famine.

In 1877, a total of 17 provinces were involved in the allocation of money and grain (Figure 5). The main drought-stricken areas, Shanxi, Henan, Shandong, Shaanxi, and Hebei, received the most amount of silver and grain. From the perspective of spatial pattern, the allocation is featured with a core-ring structure: (1) Shanxi, lying in the core, received the most amount of silver and grain; (2) Shandong, Henan, and Hebei, lying on the second layer, received the relief from other provinces and also supplied resources for Shanxi; (3) Hunan, Anhui, Jiangxi, Jiangsu and other provinces in the Middle–Lower Yangtze Plain, lying on the most peripheral ring, experienced mainly the outflow of silver and grain. In 1877, the straight-line distances of silver and grain allocations were between 270 and 1635 km, and the average transfer distance was approximately 800 km.

In 1878, also 17 provinces were involved in the relief allocations, but the spatial structure appeared to be a complex network (Figure 5). The amounts of the relief grain being allocated to Shanxi, Henan and Hebei accounted respectively for 34%, 39% and 27% of the total number of grain transfers. The distribution appeared to be more balanced than that in 1877. Shanxi, Shaanxi, Henan, and Hebei were the main receivers of the relief silver, accounting for 96% of the total silver transfers. Zhejiang, Hubei, and Hunan in the Yangtze River Basin remained as the suppliers. Apart from them, Fujian, Guangdong and Sichuan in the farther south also became the main relief suppliers. In 1878, the straight-line distances of silver and grain allocations were 132–1635 km, with an average of approximately 860 km.

Unlike the spontaneous migrations of famine victims, the allocation of money and grain is government action. It was at a larger spatial scale and with more frequent transfers. Moreover, the spatial pattern of the money and grain allocations was relatively more complex, as it would vary according to the severity of the disasters and national relief policies.

3.1.3. Temporal Characteristics of Famine-Related Migrations and Money and Grain Allocations

The records of the money and grain allocations mainly came from official documents, so the time of the documents was the time of the allocations. While the records of famine-related migrations might only appear when there were many migrations, they can still roughly indicate the period of mass famine-related migration.

According to the records that contain clear time information of the events in 1877–1878 (Figure 6), the relief allocations and famine-related migrations appeared to be seasonal and temporary, and shared with a similar peak period, which was from October 1877 to May 1878. However, the duration of money and grain allocations was longer than that of famine-related migrations.

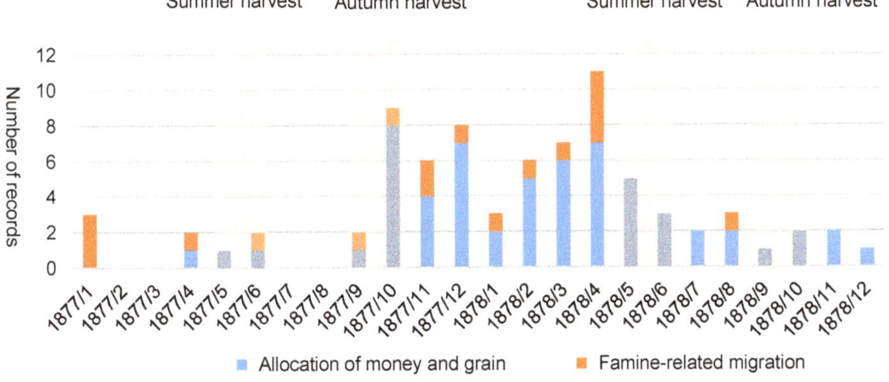

Figure 6. The number of the records of famine-related migrations and relief allocations in 1877–1878.

In North China, during the Qing dynasty, the summer harvest season started from May to June, and the autumn harvest season was from September to October. For example, in Shandong Province, wheat was harvested in the fifth lunar month (June in the Gregorian calendar), while sorghum and millet were harvested in the eighth lunar month (September in the Gregorian calendar) [49].

A small number of records of migrations before the summer harvest in 1877 indicated the occurrence of drought, but it also suggested that the situation was not beyond control. However, a large number of records were found of the time period from October 1877 to May 1878. The autumn of 1877 experienced a severe harvest failure as the consequence of persistent extreme droughts in the previous months. In the spring and summer of 1878, as the secondary disaster, plagues began to spread in the disaster areas. These two might be the main causes of the significant rise of the number of records, which suggested that the impacts of the extreme droughts had exceeded the response capability of a single region. In this stage, regional interactions were necessary for the mitigation of disaster impacts.

4. Discussion

4.1. Influence of Money and Grain Allocations on Regional Food Prices

Food prices directly reflect the food supply and demand, and are negatively correlated with the crop yields, and are also the indicators of social stability [50]. Therefore, variations in food prices in disaster areas and non-disaster areas can indicate the effects of regional interactions and coordinated responses to the extreme drought events. Wheat prices from 1876 to 1879 were selected from the Food Price Database in the Qing Dynasty [41] for the analysis, as wheat has always been the primary food staple in China. Figure 7 shows the lowest and highest wheat prices in different provinces and regions in each year from 1876 to 1879.

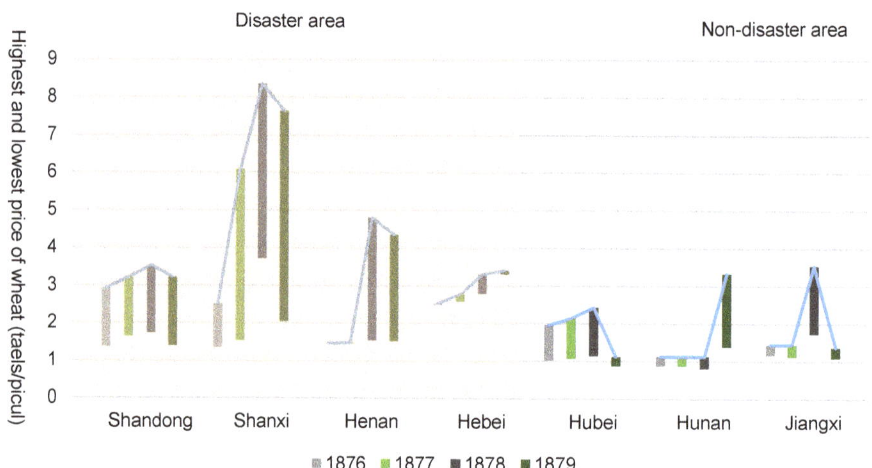

Figure 7. Provincial wheat prices in 1876–1879. Blue lines: variations of the highest annual wheat price in the province.

The wheat prices in every province of the disaster area had risen with fluctuations. In Shanxi and Henan, the prices in 1877 and 1878 could be over four or five times higher than usual. Wheat prices in Shanxi fluctuated most violently, and the extent of the rise was the most significant. In 1878 and 1879, wheat prices in Henan soared. In Shandong and Hebei, however, the wheat prices were relatively stable, with just small increases. A positive correlation can be observed between wheat price variations and the severity of droughts and famines.

The relief silver and grain alleviated some of the damages caused by the famine, but the effects were still insufficient. Taking Shanxi as an example, from 1876 to 1879, Shanxi had received 1.76 million piculs of grain and 13 million taels of silver in total. The relief silver could purchase about 2.6 million piculs of wheat according to the average annual wheat price (5 taels of silver per picul) in Shanxi at that time. That is to say, on average, about 1.09 million piculs of wheat were distributed to Shanxi each year. This amount of wheat could support 650,000 famine victims to survive for 8 months after the autumn harvest failure till the next summer harvest, based on the average daily food consumption of approximately 0.007 piculs per person [29]. However, at that time, the number of famine victims in Shanxi had exceeded 4 million [51], which means only 16% of the famished people could be supported. The situation remained severe, and some starving people even started resorting to cannibalism for survival. According to the records, starvation cannibalism occurred in respectively 42 worst-hit counties of Shanxi, 21 in Henan, and 11 Shaanxi [52].

In the non-disaster area, there were also obvious fluctuations in wheat prices. During 1877–1878, the harvest rates in Hubei, Hunan, and Jiangxi were just around 60–70% (Table 3). Sending relief grain to other provinces caused insufficient domestic supply, resulting in the sharp increases in wheat prices in Hunan and Jiangxi. On the one hand, inter-regional grain transfers worked effectively in stabilizing food prices in the disaster area. On the other hand, to some extent, the grain transfers disturbed the conditions of the food markets in the non-disaster area.

4.2. Influence of Famine-Related Migrations on Regional Social Stability

The regional interactions in social responses to the famine also profoundly influenced social stability. In 1877, more social unrest events ($\geq$2–4/10,000 km^2) took place in Henan, Shandong, Hebei, and Beijing (Figure 8), which were the main origins and destinations of the starving migrants. Very high unrest events density ($\geq$6–7/10,000 km^2) was found in Hebei. There were many reports of banditry and food robbery at the junction of Shaanxi, Shanxi, and Henan provinces. In 1878, the popular destination of famine migrants changed from the southern Jiangsu to Anhui, which made the social order of the former begin to improve. The places with a high density of social unrest events ($\geq$2–4/10,000 km^2) were recognized in Henan.

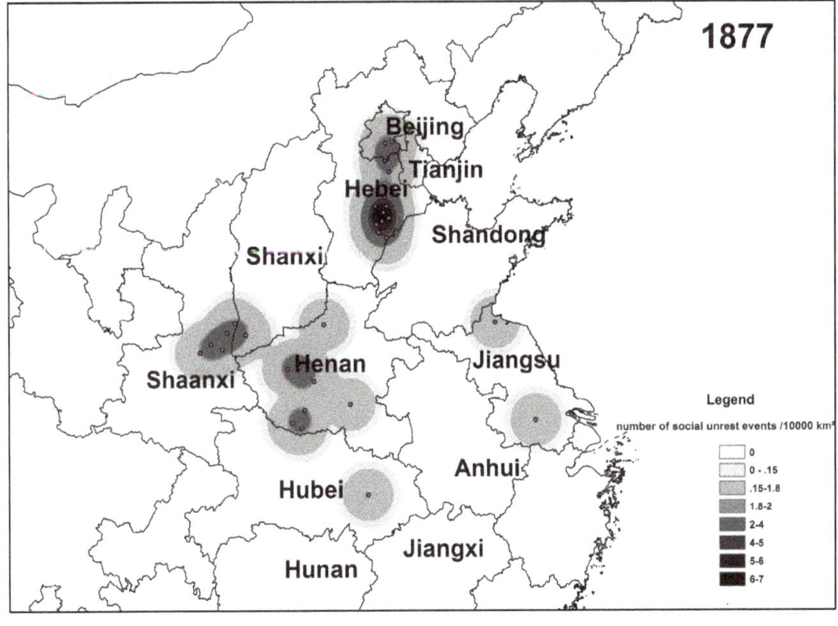

(a)

Figure 8. *Cont.*

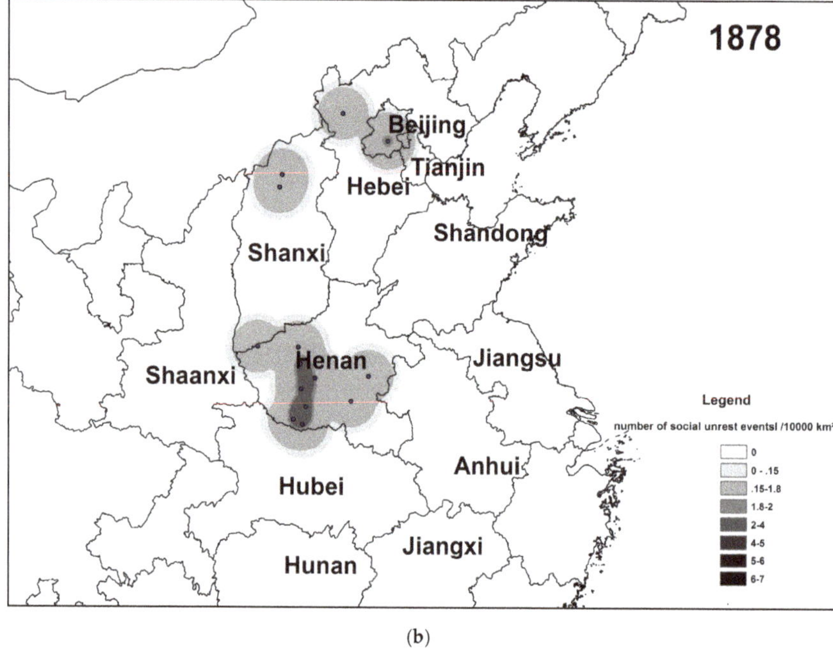

(b)

Figure 8. (**a**) Social unrest events density in 1877. (**b**) Social unrest events density in 1878.

When the famine victims left from the worst-hit areas to the surrounding slightly-impacted areas, or from villages to towns and cities, the social unrest events also "followed" with them from the disaster areas to the destinations for migrants. Thus, for the non-disaster areas, the management of the starving migrants was a challenge of regional governance.

4.3. Regional Interaction Responses and Transfer-Dispersion of the Impacts of Extreme Weather Events

Famine-related migration represents the dispersion of the population pressure in the famine-struck areas (Figure 9). The spatial redistribution of the famine victims dispersed the population pressure in their places of origin, but increased the population pressure in their destinations. Famine-related migration is also connected to social unrest events and represents the dispersion of the social impacts of the extreme drought events.

The allocation of money and grain is an administrative action, with the purpose of adjusting the gaps in food production between disaster areas and non-disaster areas, and increase the opportunities to obtain food for the victims in the disaster areas. Affected by the allocation, food prices in some of the non-disaster areas also increased. To a certain extent, the allocation of money and grain played a role in transferring the social impacts of disasters and had positive effects on the post-disaster recovery (Figure 9).

In Chinese history, normally there were two areas of destination for the migration driven by the events related to climate change. One is the southeastern part of China, with warm and humid weather and fertile soils. The other is the arid and barren northwest. Migrating to the northwest seems to go against common sense, but it was because of the increase of conflicts between farmers and herders, and the increase of invasion of the northern nomads in the dry period [53]. Between 1876 and 1879, extreme droughts led to a server famine in North China, and many people migrated in search of food. However, in spite of the food production, the starving migrants selected their destinations using the location and proximity to their homeland as the determining factors. This is because that the famine-related migrations at that time were mainly temporary, and it was difficult for starving people to complete long-distance travel. Compared with the climate-driven migration mentioned above, the migration during the North China Famine of 1876–1879 was a spontaneous social response to the extreme disaster events and the dispersion of short-term population pressure in the disaster areas.

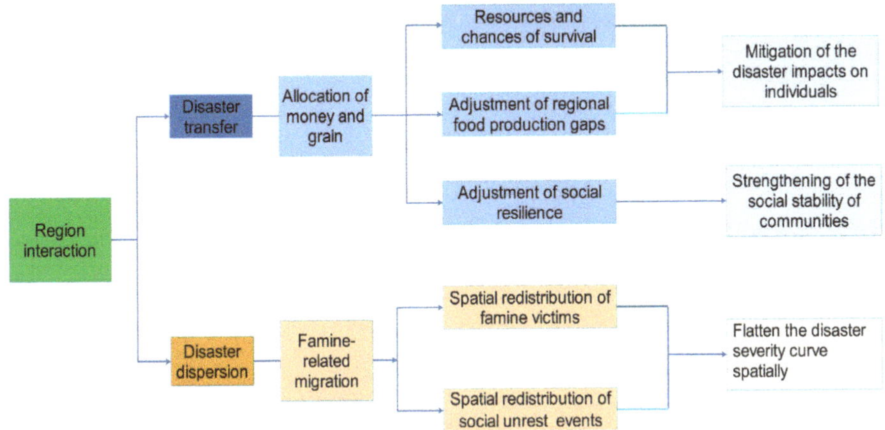

Figure 9. Process of regional interaction responses and transfer-dispersion of the impacts of disasters.

Based on the vast canal network, the central government planned, organized and carried out the allocation of money and grain, which was also the process of impact transfer across the whole country. In this network, the relationship between regions was more complex than that in the migration network. This time, in spite of the distances, the grain storage and accessibility of a place were the main factors in the decision-making process.

5. Conclusions

After the analysis of the spatial and temporal characteristics of famine-related migration and allocation of the disaster relief in the North China Famine of 1876–1879, the conclusions were summarized as follows:

(1) Famine caused by extreme drought events was the main driving force of the migration. Famine-related migration was spontaneous and short-distanced, flowing into the surrounding towns and cities. The straight-line travel distances of most migrations were approximately 400 km. Famine-related migration spatially dispersed the population pressure but caused the spillover of social unrest.

(2) As a government action, the relief silver and grain from the non-disaster areas were distributed to the disaster areas, with an average relief transfer distance of over 800 km. The transfers of the famine relief formed a complex spatial network. During the worst period of the famine, due to harvest failures, wheat prices were over four or five times higher than usual. The allocation of money and grain relieved some pressure on the food supply in the disaster areas but did not fundamentally change the situation. It also affected the equilibrium of the food market in the non-disaster areas, which led to the fluctuations in wheat prices.

(3) The regional interactions in the process of responding to extreme climate events is a process of dispersion and transfer of the disaster events' impacts, which will have different risk effects to both disaster areas and non-disaster areas. In the context of increasing globalization and regional linkages, a higher capacity for integrated risk prevention and comprehensive administrative governance is required.

Author Contributions: X.Z. performed research, analyzed data, and wrote the original manuscript; X.F. revised the original manuscript of the paper and made contributions to the research ideas. Y.S. designed research, revised the original manuscript of the paper and discussed the results. All authors have read and agreed to the published version of the manuscript.

Funding: This research was funded by National Key Research and Development Program of China (No. 2018YFA0605602) and National Natural Sciences Foundation of China (No. 41771572).

Conflicts of Interest: The authors declare no conflict of interest.

References

1. Diaz, H.F.; Stahle, D.W. Climate and cultural history in the Americas: An overview. *Clim. Chang.* **2007**, *83*, 1–8. [CrossRef]
2. Costanza, R.; van der Leeuw, S.; Hibbard, K.; Aulenbach, S.; Brewer, S.; Burek, M.; Cornell, S.; Crumley, C.; Dearing, J.; Folke, C.; et al. Developing an integrated history and future of people on earth (IHOPE). *Curr. Opin. Environ. Sustain.* **2012**, *4*, 106–114. [CrossRef]
3. Webb, P. Emergency relief during Europe's famine of 1817 anticipated crisis-response mechanisms of today. *J. Nutr.* **2002**, *132*, 2092S–2095S. [CrossRef]
4. O'Gorman, E.; Beattie, J.; Henry, M. Histories of climate, science, and colonization in Australia and New Zealand, 1800–1945. *WIREs Clim. Chang.* **2016**, *7*, 893–909. [CrossRef]
5. MacDonald, G.M. Severe and sustained drought in southern California and the west: Present conditions and insights from the past on causes and impacts. *Quat. Int.* **2007**, *173*, 87–100. [CrossRef]
6. Pei, Q. Migration for survival under natural disasters: A reluctant and passive choice for agriculturalists in historical China. *Sci. China Earth Sci.* **2017**, *60*, 2089–2096. [CrossRef]
7. Chi, Z.-H.; Li, H.-Y. Disasters, social changes and refugees—A focus on modern Hebei at the turn of the 20th century. *J. Nanjing Agric. Univ. (Soc. Sci. Ed.)* **2004**, *4*, 70–78. [CrossRef]
8. Xiao, J.; Zheng, G.; Guo, Z.; Yan, L. Climate change and social response during the heyday of the little ice age in the Ming and Qing dynasties. *J. Arid Land Resour. Environ.* **2018**, *32*, 79–84. [CrossRef]
9. Xiao, L.; Fang, X.; Zheng, J.; Zhao, W. Famine, migration and war: Comparison of climate change impacts and social responses in North China between the late Ming and late Qing dynasties. *Holocene* **2015**, *25*, 900–910. [CrossRef]
10. Xiao, L.; Huang, H.; Wei, Z. Comparison of governmental relief food scheduling and social consequences during droughts of 1743–1744 AD and 1876–1878 AD in North China. *J. Catastrophol.* **2012**, *27*, 101–106. [CrossRef]
11. Huang, H.; Xiao, L.-B.; Luo, Y.-H.; Fang, X.-Q. Temporal and spatial variation of the amount of relief grain scheduling in the North China plain in the Qing Dynasty. *J. Earth Environ.* **2014**, *5*, 410–416. [CrossRef]
12. Shiue, C.H. The political economy of famine relief in China, 1740–1820. *J. Interdiscip. Hist.* **2005**, *36*, 33–55. [CrossRef]
13. Costanza, R.; Graumlich, L.; Steffen, W.; Crumley, C.; Dearing, J.; Hibbard, K.; Leemans, R.; Redman, C.; Schimel, D. Sustainability or collapse: What can we learn from integrating the history of humans and the rest of nature? *AMBIO A J. Hum. Environ.* **2007**, *36*, 522–527. [CrossRef]
14. Herring, S.C.; Christidis, N.; Hoell, A.; Hoerling, M.P.; Stott, P.A. Explaining extreme events of 2018 from a climate perspective. *Bull. Am. Meteorol. Soc.* **2020**, *101*, S1–S128. [CrossRef]
15. Field, C.B.; Barros, V.; Stocker, T.F. *Managing the Risks of Extreme Events and Disasters to Advance Climate Change Adaptation: Special Report of the Intergovernmental Panel on Climate Change;* Cambridge University Press: Cambridge, UK; New York, NY, USA, 2012; p. 6, ISBN 1107025060.
16. Alexander, D. Globalization of disaster: Trends, problems and dilemmas. *J. Int. Aff.* **2006**, *59*, 1–22.
17. Ye, T.; Shi, P.; Wang, J.A.; Liu, L.; Fan, Y.; Hu, J. China's drought disaster risk management: Perspective of severe droughts in 2009–2010. *Int. J. Disaster Risk Sci.* **2012**, *3*, 84–97. [CrossRef]
18. Brázdil, R.; Kiss, A.; Luterbacher, J.; Nash, D.J.; Řezníčková, L. Documentary data and the study of past droughts: A global state of the art. *Clim. Past* **2018**, *14*, 1915–1960. [CrossRef]
19. Nash, D.J.; Pribyl, K.; Klein, J.; Neukom, R.; Endfield, G.H.; Adamson, G.C.D.; Kniveton, D.R. Seasonal rainfall variability in southeast Africa during the nineteenth century reconstructed from documentary sources. *Clim. Chang.* **2016**, *134*, 605–619. [CrossRef]

20. Añel, J.A.; Sáenz, G.; Ramírez-González, I.A.; Polychroniadou, E.; Vidal-Mina, R.; Gimeno, L.; de La Torre, L. Obtaining meteorological data from historical newspapers: La integridad. *Weather* **2017**, *72*, 366–371. [CrossRef]

21. Noone, S.; Broderick, C.; Duffy, C.; Matthews, T.; Wilby, R.L.; Murphy, C. A 250-year drought catalogue for the island of Ireland (1765–2015). *Int. J. Clim.* **2017**, *37*, 239–254. [CrossRef]

22. Zheng, J.; Hao, Z.; Fang, X.; Quansheng, G.E.; Zhi, L.I.; Liu, W.; Zheng, F. Changing characteristics of extreme climate events during past 2000 years in China. *Prog. Geogr.* **2014**, *33*, 3–12. [CrossRef]

23. Hao, Z.; Zheng, J.; Wu, G.; Zhang, X.; Ge, Q. 1876–1878 severe drought in North China: Facts, impacts and climatic background. *Chin. Sci. Bull.* **2010**, *55*, 3001–3007. [CrossRef]

24. Man, Z.-M. Climatic background of the severe drought in 1877. *Fudan J.* **2000**, 28–35.

25. Nicholls, N. The EI Niño/southern oscillation and Australian vegetation. *Plant* **1991**, *91*, 23–36. [CrossRef]

26. Diaz, H.F.; Markgraf, V. *El Niño: Historical and Paleoclimatic Aspects of the Southern Oscillation*; Cambridge University Press: Cambridge, UK; New York, NY, USA, 1992; pp. 151–174, ISBN 0521430429.

27. Herweijer, C.; Seager, R.; Cook, E.R. North American droughts of the mid to late nineteenth century: A history, simulation and implication for Mediaeval drought. *Holocene* **2006**, *16*, 159–171. [CrossRef]

28. Aceituno, P.; del Rosario Prieto, M.; Solari, M.E.; Martínez, A.; Poveda, G.; Falvey, M. The 1877–1878 El Niño episode: Associated impacts in South America. *Clim. Chang.* **2009**, *92*, 389–416. [CrossRef]

29. Li, L.M. *Fighting Famine in North China: State, Market, and Environmental Decline, 1690s–1990s*; Stanford University Press: Stanford, CA, USA, 2007; ISBN 0804753040.

30. Xia, M. Discuss "Ding-wu disaster". *Qing Hist. J.* **1992**, *4*, 83–91.

31. Wang, J. Analysis of the causes of drought in northern China. *Acad. J. Jinyang* **2000**, *6*, 56–61.

32. Yang, J. The changes of the relief in the late Qing Dynasty. *Qing Hist. J.* **2000**, *4*, 59–64.

33. Li, W. *Top Ten Disasters in Modern China*; Shanghai People's Publishing House: Shanghai, China, 1994; pp. 80–114, ISBN 9787208018129.

34. *Qing Shi Lu*; Zhonghua Book Company: Beijing, China, 2008; Volume 52, pp. 1–879, Volume 53, pp. 1–562; ISBN 978710105626.

35. Han Tang Modern Newspaper Database. Shenbao. Available online: https://www.neohytung.com/ (accessed on 10 December 2019).

36. Li, W.; Lin, D.; Cheng, X.; Gong, M. *Disaster Annals in Recent China*; Hunan Education Press: Changsha, China, 1990; pp. 341–415, ISBN 9787535510839.

37. Dai, Y. *Qing Tong Jian*; Shanxi People's Publishing House: Taiyuan, China, 1999; Volume 17, ISBN 9787203039075.

38. Wagner Rudolf, G. The Shenbao in crisis: The international environment and the conflict between Guo Songtao and the Shenbao. *Late Imp. China* **1999**, *20*, 107–143. [CrossRef]

39. Xia, Y.; Diao, A. The "residents' concern" in the "Shenbao" of the famine report: A case study of the report of Ding-wu disaster. *News World* **2010**, *6*, 102–103.

40. Su, Q.; Zheng, W. A summary of Li Wenhai and the research of disasters of China in modern times. *J. Inst. Disaster Prev. Sci. Technol.* **2008**, *10*, 100–105. [CrossRef]

41. Food Price Database in the Qing Dynasty. Available online: http://mhdb.mh.sinica.edu.tw/foodprice/ (accessed on 10 January 2020).

42. Zhang, D.; Liang, Y. A long lasting and extensive drought event over China in 1876–1878. *Adv. Clim. Chang. Res.* **2010**, *1*, 91–99. [CrossRef]

43. Zhang, D. *A Compendium of Chinese Meteorological Records of the Last 3000 Years*; Jiangsu Education Press: Nanjing, China, 2004; Volume 4, pp. 2673–3666, ISBN 7806434682. (In Chinese)

44. Ti, L. An estimation of China's GDP from 1600 to 1840. *Econ. Res. J.* **2009**, *10*, 144–145.

45. Silverman, B.W. *Density Estimation for Statistics and Data Analysis*; CRC Press: Boca Raton, FL, USA, 1986; p. 76, ISBN 0412246201.

46. Xiao, L. Spatio-temporal distribution of natural disasters in China during 1644–1911 based on kernel density estimation. *J. Catastrophology* **2019**, *34*, 92–99. [CrossRef]

47. Xia, M. China's famine and population migration during the Anti Japanese War. *J. Stud. China's Resist. War Jpn.* **2000**, 61–80.

48. Li, W. *Materials of Modern Agricultural History of China*; Joint Publishing Company: Beijing, China, 1957; Volume 1, pp. 756–759.

49. Yin, C.M.U. Preliminary investigation on seasonal variation of the price of wheat in Shandong province during the reign of Dao Guang. *Agric. Hist. China* **2013**, *32*, 73–84.
50. Bellemare, M.F. Rising food prices, food price volatility, and social unrest. *Am. J. Agric. Econ.* **2015**, *97*, 1–21. [CrossRef]
51. Ping, H. The population loss resulting from the great drought in Shanxi in the early period of emperor Guangxu's reign in the Qing dynasty. *J. Shanxi Univ. (Philos. Soc. Sci.)* **2004**, *24*, 10–13. [CrossRef]
52. Wen, Y.; Fang, X.; Li, Y.; Xia, L. Research literature review on the great drought disaster in north China during 1876–1879. *J. Catastrophol.* **2019**, *34*, 172–180. [CrossRef]
53. Pei, Q.; Nowak, Z.; Li, G.; Xu, C.; Chan, W.K. The strange flight of the peacock: Farmers' atypical northwesterly migration from central China, 200 BC–1400 AD. *Ann. Am. Assoc. Geogr.* **2019**, *109*, 1583–1596. [CrossRef]

Article

Examining the Direct and Indirect Effects of Climatic Variables on Plague Dynamics

Ricci P.H. Yue [*,†] **and Harry F. Lee** [*,†]

Department of Geography and Resource Management, The Chinese University of Hong Kong, Shatin, New Territories, Hong Kong
* Correspondence: ricciyue@cuhk.edu.hk (R.P.H.Y.); harrylee@cuhk.edu.hk (H.F.L.)
† These authors contributed equally to the work.

Received: 15 March 2020; Accepted: 10 April 2020; Published: 15 April 2020

Abstract: Climate change can influence infectious disease dynamics both directly, by affecting the disease ecology, and indirectly, through altering economic systems. However, despite that climate-driven human plague dynamics have been extensively studied in recent years, little is known about the relative importance of the direct and indirect effects of climate change on plague outbreak. By using Structural Equation Modeling, we estimated the direct influence of climate change on human plague dynamics and the impact of climate-driven economic change on human plague outbreak. After studying human plague outbreak in Europe during AD1347–1760, we detected no direct climatic effect on plague dynamics; instead, all of the climatic impacts on plague dynamics were indirect, and were operationalized via economic changes. Through a series of sensitivity checks, we further proved that temperature-induced economic changes drove plague dynamics during cold and wet periods, while precipitation-induced economic changes drove plague dynamics during the cold periods. Our results suggest that we should not dismiss the role of economic systems when examining how climate change altered plague dynamics in human history.

Keywords: climate change; plague; direct and indirect effects; Structural Equation Modelling

1. Introduction

Arguably, there is an increasing consensus in academia showing that climate change is a dominant driving force of plague outbreak in the intricate pattern of disease complexity. For example, climatic fluctuations have been found associated with the recurrent introduction of *Yersinia pestis*, the bacterium responsible for causing the plague, from Asia into Europe during the Black Death era [1,2]. Stenseth, et al. [3] pointed out that warmer springs and wetter summers favored the prevalence of plague dynamics at its natural reservoir in Kazakhstan. In China, the modeling results by Xu, et al. [4] demonstrated that historical plague intensities in Northern and Southern China were positively correlated to wetness and dryness, respectively. However, according to more updated findings from Xu, et al. [5], wetness accelerated the spread of the plague during the third plague pandemic in China. Apparently, some of these climate-plague nexuses are related to climatic phenomena at a larger scale. Plague outbreak is quantitatively known to be associated with El Nino Southern Oscillation [6–8], Indian Ocean Dipole [7], Pacific Decadal Oscillation [8], and Southern Oscillation Index [9].

Although the above studies help us understand the connection between climate change and plague outbreaks, they are all grounded on the same assumption: climatic variables have a direct and one-step impact on plague dynamics. In those studies, the investigated climatic variables and the plague-indicating variables are simply correlated for measuring their first-order relationship. However, is the association between climate and plague straightforward in nature? According to a series of studies by Zhang et al. [10–12], climatic variations might have worked through a specific set of economic systems to cause the outbreak of epidemics. Hence, the climate-plague association is unlikely to be direct in nature, as there should be a medium in human societies translating climatic forcing to plague outbreaks.

If there is indeed a direct relationship between the climatic variables and plague dynamics, then the investigations of the climate-plague nexus would remain convincing that the nexus normally requires a lesser likelihood to include a third variable in explaining the correlation. If there is not such a relationship, although it cannot overturn the previous observations and computational results, it may somehow indicate that many scholars may have oversimplified the necessary pathways between climate change and plague outbreak. As Zhang et al. [12] indicated, climate change could generate chronic undernourishment in China through the shrinkage of agricultural production, which further weakened immunity to infectious diseases. Lee and Zhang [13] associated the causality of climate-induced famine on epidemics outbreak in historical China. Tian, et al. [14] worked further to reveal that cold and dry climate conditions, in large part, indirectly increased the frequency of epidemics outbreak through generating the prevalence of locusts and famines in China over the last two millennia. Dunca and Scott [15] also emphasized the crucial role of nutrition and immunity in the spread of infectious diseases in pre-industrial European societies. The study by Pei, et al. [16] outlined the dependence on climate of the macroeconomic structure in Europe, and further indicated that such dependence would be imperative in pulling human societies into the Malthusian trap, which includes epidemics outbreak, in pre-industrial Europe. Specifically, we also provided a hypothetical explanation of the climate-plague relationship and highlighted the possibility that climate change affected plague dynamics via economic systems. Hence, the influence of temperature/precipitation on plague dynamics appeared to have inconsistent time lags at the multi-decadal to centennial timescales [2]. However, despite efforts to demonstrate the climate-economic-epidemics relationship in the past decades, there exists little empirical understanding of how climatic variations and plague dynamics are integrated via economic systems.

In this study, we tried to fill in the research gap in the climate-economic-epidemic relationship by examining the direct and indirect effects of climatic fluctuations on the frequency of plague outbreak in pre-industrial Europe at the continental scale. In our study region over the course of AD1347–1760, there have been documented long-term counts of human plague outbreak; the economic trend has been described in fine resolution; and climatic variables have been reconstructed through widespread dendrochronology records. We used such long-term data on plague, climate, and economic attributes to answer the following questions: (1) whether the investigated climatic variable exhibited a direct and/or indirect influence (through economic systems) on human plague dynamics in Europe at the continental scale; (2) whether economic attributes deliver their influence to human plague activities in Europe at the continental scale; and (3) whether both climatic variables and economic factors maintain a long-term trend with human plague dynamics over the study period in Europe at the continental scale. To address the above issues, we applied structural equation modeling (SEM) to disentangle that direct and/or indirect importance of climatic fluctuations on historical plague activities in human societies. We showed all the hypothetical pathways being tested in our analysis in Figure 1. Our findings suggest that climate-driven economic fluctuations played a crucial role in translating climate change into plague outbreak.

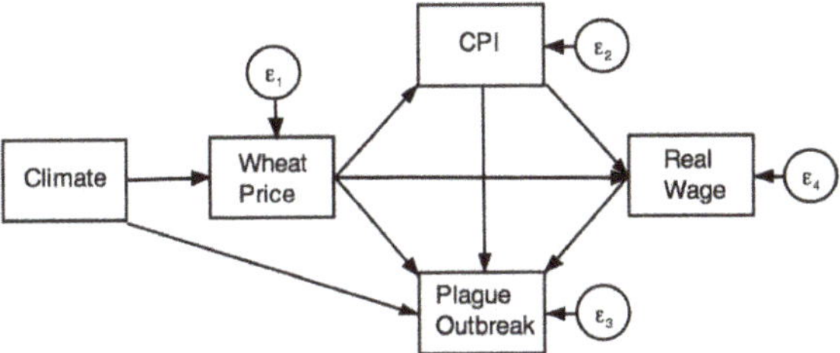

Figure 1. Path diagram showing all hypothesized direct and indirect links amongst climate change, economic fluctuations (including wheat price, CPI, and real wage), and plague outbreak. Climate change and economic fluctuations may directly cause the plague outbreak. Also, climate change may indirectly cause the plague outbreak by influencing the wheat price, and then the wheat price will affect CPI and the real wage.

2. Materials and Methods

2.1. Plague Data

To measure the historical plague outbreak in Europe, we adopted a detailed geo-referenced plague dataset digitalized by Büntgen, et al. [17]. The database records the starting year of each human plague outbreak in Europe at the city-scale. We counted the number of cities identified with plague outbreak in each year and transformed the database into a time series. Only data from Europe was counted, as Europe was selected for our study region. Slightly different from the original dataset, which used the number of total plague outbreak count as its unit, the transformed dataset adopted in this study used the number of cities with plague outbreak count as our unit. Over the AD1347–1760 period, a total of 6764 plague outbreaks were recorded in Europe. The year with the most extensive plague outbreak is AD1630, with a record 119 cities affected by the plague. Out of 414 units of observation, plague quiet years are recorded 14 times. Zero-inflation of the dataset should not be considered as a problem.

2.2. Climate Data

We considered three sets of climate data, namely temperature, precipitation, and scPDSI, for our analysis in this study. They were common parameters for historical study relating to climate-human relationships and were found strongly significant in influencing plague dynamics over time and space [18,19].

The temperature and precipitation dataset included for analysis are acquired from the climate reconstruction provided by Büntgen, et al. [20]. This climatic reconstruction was made possible through surveying 1547 sets of tree ring chronologies from Europe for the reconstruction of past climatic variability of Europe at its continental scale in annual resolution. Particularly, the time series for temperature data is calibrated into temperature anomaly with respect to the period of AD1901–2000. The vast coverage of raw data from this dataset ensured the reliability and validity of climate reconstruction. Thus, the dataset has been widely adopted in other historical studies of Europe [21].

Another set of climatic variables adopted in this study originated from the Old World Drought Atlas (OWDA) project of Cook et al. [22]. Our study considered self-calibrated Palmer Drought Severity Index (scPDSI) as the projection of the natural hydroclimatic environment. As such, the PDSI reconstruction by Cook et al. [22] provides an ideal extended record of natural wetness/dryness variability for the pre-industrial era of Europe. The OWDA was developed from dendrochronological records over the European continent and calibrated with high-quality instrumental scPDSI gridded data from the Royal Netherlands Meteorological Institute. Our study retrieved available data points, which have a spatial resolution of half-degree longitude-by-latitude grid, from our study area and further aggregated relevant grids of the same year into a time series for our analysis.

2.3. Economic Data

For historical economic parameters, we selected wheat price, consumer price index (CPI), and real wages as variables for our estimations. The historical wheat price is extracted from the database created by Allen [23]. We extracted the data from each city and determined whether they fell onto our study region. The raw data is first standardized by $[(x_i - x_{mean})/x_{s.d.}]$, then we calculated the averaged standardized wheat price of Europe. The historical CPI data comes from Allen [23]. We calculated the averaged standardized CPI by the same method as suggested in wheat price. We here averaged the standardized real wages of laborers and standardized real wages of building craftsmen based on the database from Allen [23].

2.4. Structural Equation Modeling

We applied structural equation modeling (SEM) [24] to test for the relative importance of climatic variables and economic variables on plague dynamics, and whether climate change has a direct influence on plague dynamics. To do this, we first constructed all the hypothetical pathways that fully detailed the causality amongst variables within the system being studied [25]. Then, mathematically, the total pathway added up together as a series of linear regression. The technique hypothetically decomposed all the correlations of two variables into direct effects that pinpointed the causal influence of one factor on another and indirect effects that passed through other variables in the model and non-casual mediating predictors resulting from a common cause [26]. From the full casual model, the sum of direct effects and all indirect effects mediated by other variables between the predictor and response variable would yield total effect. In SEM, by assuming that all of the important variables and pathways were labeled, every effect listed in path analysis was considered linear, additive and unidirectional, and that residuals were presumably uncorrelated [27]. In this study, as also shown in Figure 1, the response variable is plague outbreak. We hypothesized that climate change and economic fluctuations (including wheat price, CPI, and real wage) may directly cause plague outbreak. Also, climate change may indirectly cause plague outbreak by influencing wheat price, which then will affect CPI and real wage.

2.5. Linear Regressions

To test whether different sets of data correlate directionally with each other over the long-run and in different climatic settings, we statistically compared the long-term trends between: (1) climate change and plague outbreak; (2) economic change and plague outbreak; (3) climate change and economic change; and (4) internal dynamics of economic change. For each set of correlation tested, we divided the study timespan into different climatic periods, namely: (1) warm periods (positive temperature anomaly); (2) cold periods (negative temperature anomaly); (3) dry periods (below-average precipitation with reference to AD1347–1760); and (4) wet periods (above-average precipitation with reference to AD1347–1760). To assess the afore-mentioned long-term relationship, we adopted simple linear regression models for each combination.

3. Results

3.1. Direct and Indirect Climatic Effect on Plague Outbreak

Over our study period in AD1347–1760, our SEM results showed that temperature and precipitation variations would have a substantial indirect impact on plague outbreak in Europe through climate-induced fluctuations in wheat prices (Figure 2). Yet, the climatic influence, as implied from the SEM results, is never directly linked to plague dynamics.

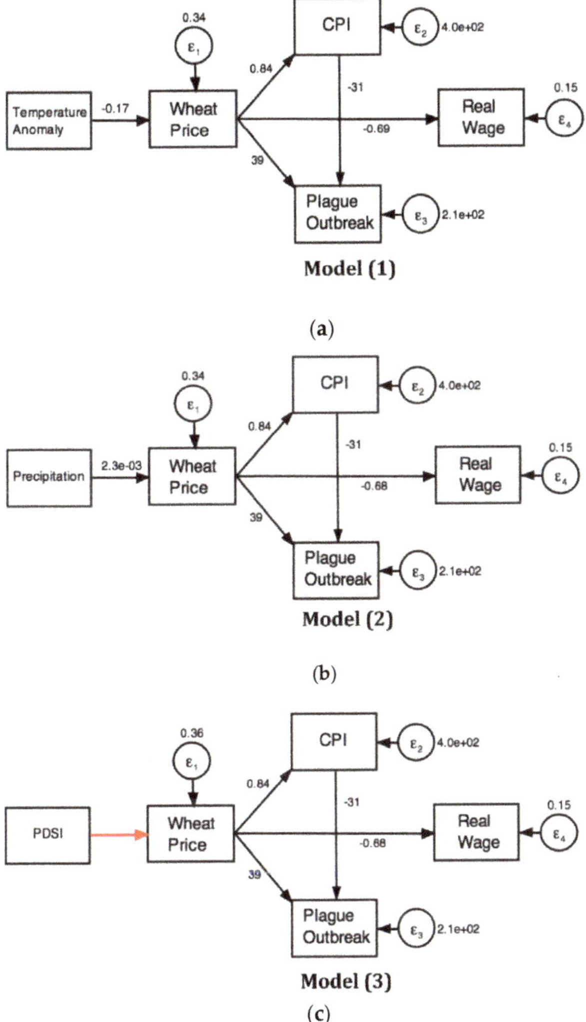

Figure 2. Path diagrams for (**a**) Model 1: Temperature anomaly; (**b**) Model 2: Precipitation; (**c**) Model 3: PDSI, for the direct and indirect effects of climate change on plague outbreak. The residual variables (ε_1, ε_2, ε_3, ε_4) represent the unmeasured factors affecting the corresponding variable. Arrows represent the relationship of each pair of the variables, with the path coefficients stated next to the arrows. The path coefficients are standardized partial regression coefficients from linear regressions. We omit those statistically insignificant paths in the models, except for the path PDSI → Wheat Price. Red arrows represent statistically insignificant paths, while the black ones represent statistically significant ones.

Model 1 shows that temperature variation has a statistically-insignificant direct relationship with plague dynamics (Table 1).

Table 1. Correlation coefficients between each set of climatic/economic predictor and plague response. The correlation is decomposed into the direct and indirect effects, and the synergy of the direct and indirect effects gives the total effect.

	Direct	Indirect	Total
Model 1			
Temp → Plague	−1.49	−1.66 ***	−3.15 ***
Wheat price → Plague	38.56 ***	−28.86 ***	9.69 **
CPI → Plague	−31.68 ***	−0.32	−31.99 ***
Real wage → Plague	3.02	/	3.02
Model 2			
Precipitation → Plague	−0.03	0.02 ***	−0.01
Wheat price → Plague	38.79 ***	−28.31 ***	1048 ***
CPI → Plague	−31.12 ***	−0.30	2.91 ***
Real wage → Plague	2.91	/	2.91
Model 3			
PDSI → Plague	−0.51	0.09	0.47
Wheat price → Plague	38.91 ***	−28.70 ***	10.21 ***
CPI → Plague	−31.43 ***	−0.32	−31.75 ***
Real wage → Plague	3.08	/	3.08

Significance level: *** $p < 0.001$, ** $p < 0.05$.

However, temperature would indirectly control the variations of plague frequency through wheat price fluctuations (intercept = −1.66, $p < 0.001$). In Model 2, precipitation displayed a similar pattern as temperature. The command of precipitation on plague dynamics is undertaken indirectly through the manipulation of wheat prices (intercept = 0.02, $p < 0.001$) during our study period. Likewise, the path model again denied the direct impact of precipitation on the plague outbreak. In Model 3, the estimation suggested that PDSI would have no directional effect on plague dynamics, both directly and indirectly. For each set of the path analysis, we also included selected economic factors to investigate the direct and indirect influence of climate change on plague dynamics. In all models, it is observed that, despite the climate-induced fluctuation, both wheat price and CPI have a direct effect on plague outbreak. The influence of wheat price and CPI was not mediated by other variables in the models when other factors were held constant. The models also universally demonstrated that the sensitivity of plague dynamics was indirectly correlated with the changes in wheat price via CPI. However, not all the economic variables tested were found relevant to plague outbreak. The historical real wage was not directly related to any variations of plague dynamics. By combining the direct effect and indirect effects of predictors, we were able to measure the total effect of both climatic variables and economic variables on plague outbreak. From the result, as indicated in Table 1, the total effect of temperature, precipitation, and CPI remained statistically negatively significant to any change of plague activity, while the effect of wheat price on plague was reported to be positive. In short, plague dynamics is favored by low temperature, dry environment, rising wheat price, and decreasing CPI.

3.2. Long-Term Trends of Climate, Economic Changes, and Plague Outbreak

In the process of creating SEM from climatic, economic, and plague data, we also looked for long-term trends between variables and checked for their consistency over our study period. In Figure 3, we laid out the general long-term sensitivity of plague dynamics and identified predictors deduced from SEM.

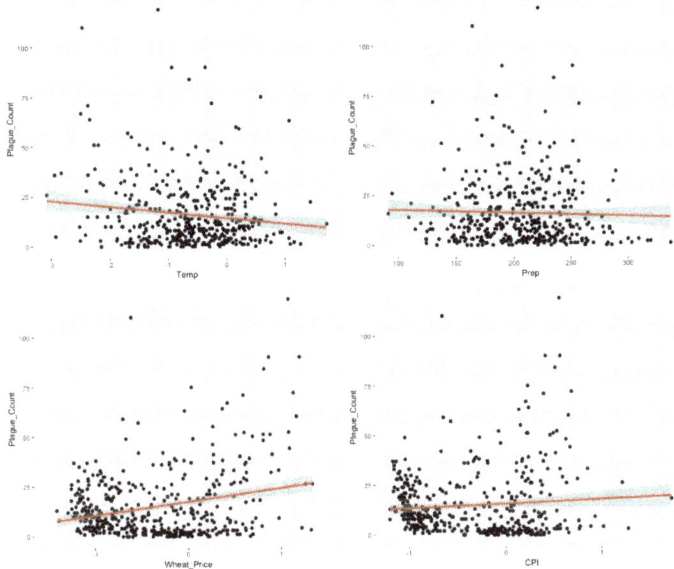

Figure 3. Long-term trend of plague outbreak with (**top left**) temperature anomaly; (**top right**) precipitation; (**bottom left**) wheat price; and (**bottom right**) CPI. The red lines represent the trends, and the green envelopes provide the 95% confidence interval areas.

Plague dynamics, as estimated, had a negative trend with temperature (Coef. $= -0.007$, $p < 0.001$, F = 8.24) (Table A1), implying that cooling would effectively trigger plague outbreak. For precipitation, the long-term trend is statistically insignificant to the plague dynamics. For the two economic variables tested here, wheat price exhibited a consistent positive trend with the plague dynamics (Coef. $= 0.0126$, $p < 0.001$, F = 60.24); whilst CPI also showed a similar trend (Coef. $= 0.0062$, $p < 0.001$, F = 16.20).

Over the 414 years of observation, temperature and precipitation both have a persistent impact on wheat price. It was estimated that the price of wheat drops with increasing temperature (Coef. $= -0.319$, $p < 0.001$, F = 23.81) and decreasing rainfall (Coef. $= 10.519$, $p < 0.001$, F = 23.81). At the same time, wheat price, because it is closely related to the economy, was itself positively correlated with CPI (Coef. $= 1.029$, $p < 0.001$, F = 2597.16).

We also compared the long-term trend of temperature influence between warm and cold periods. It should be noted that climatic control on plague and economic parameters behaved differently in warm and cold periods. During cold periods, the long-term trend of all studied relationships performed the same as the overall long-term trend observed (Figure 4, Table A2). However, during warm periods, temperature no longer exerted its effect on plague dynamics and wheat price (Figure 5, Table A3). From the statistical results we obtained, the sensitivity of plague dynamics was primarily controlled by wheat price (Coef. $= 0.0122$, $p < 0.001$, F = 12.04) and CPI (Coef. $= 0.0093$, $p < 0.001$, F = 8.66) within the warm phases. Such control was also revealed in the SEM models presented in the previous section, in which growing CPI and rising wheat price occurred together during the warm periods (Coef. $= 1.065$, $p < 0.001$, F = 691.12).

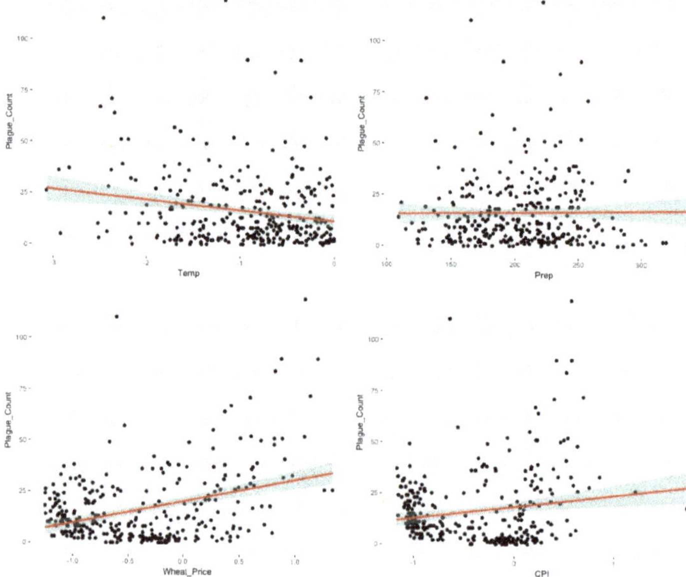

Figure 4. Long-term trend of plague outbreak with (**top left**) temperature anomaly; (**top right**) precipitation; (**bottom left**) wheat price; and (**bottom right**) CPI during the cold periods. The cold periods refer to the time with negative temperature anomaly. Red lines represent the trends, and the green envelopes provide the 95% confidence interval areas.

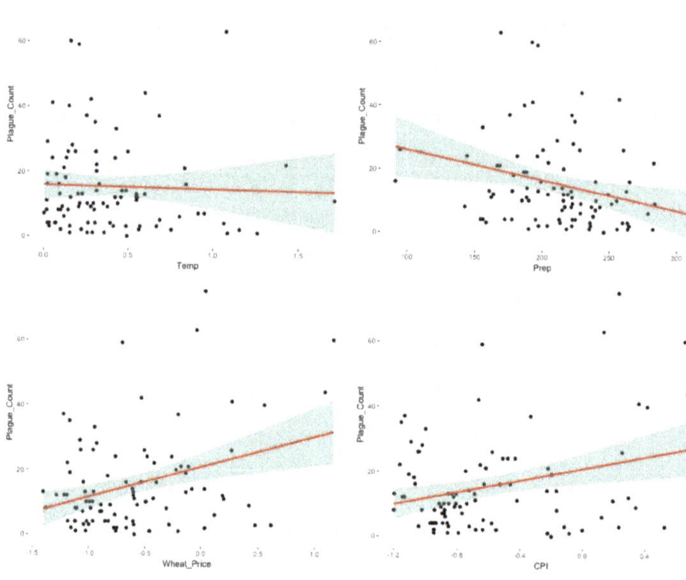

Figure 5. Long-term trend of plague outbreak with (**top left**) temperature anomaly; (**top right**) precipitation; (**bottom left**) wheat price; and (**bottom right**) CPI during the warm periods. The warm periods refer to the time with positive temperature anomaly. Red lines represent the trends, and green envelopes provide the 95% confidence interval areas.

In addition, the performance of the climatic variable differed in the dry and wet periods. Further analyses showed that temperature could not change the trend of plague outbreak during dry periods (Figure 6, Table A4). The pressure from the temperature on wheat price and plague outbreak did not exist at all during dry periods. However, in wet periods, the dynamics of plague activity increased with temperature cooling (Coef. = −0.0119, $p < 0.001$, F = 12.90), whilst increasing wheat price was also associated with decreasing temperature (Coef. = −0.524, $p < 0.001$, F = 32.18) (Figure 7, Table A5).

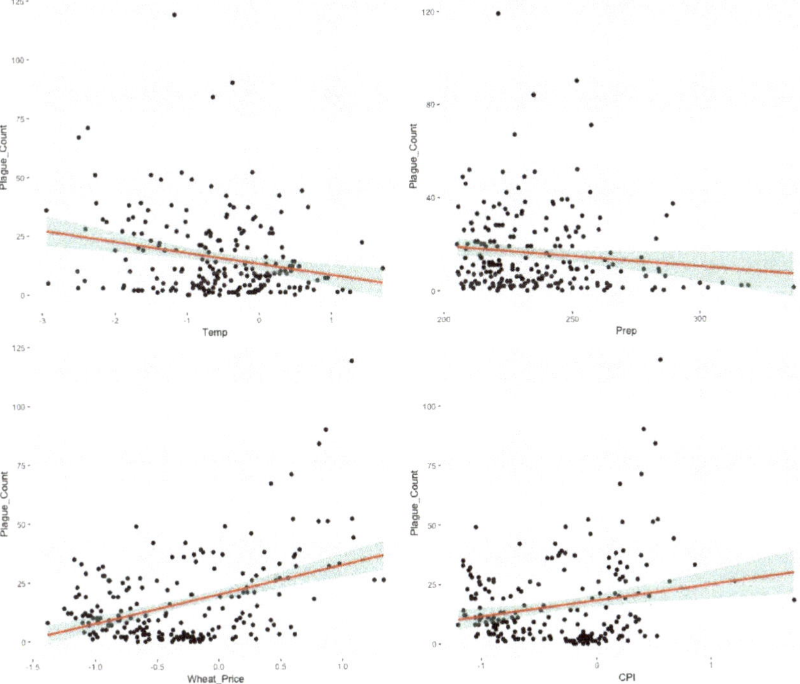

Figure 6. Long-term trend of plague outbreak with (**top left**) temperature anomaly; (**top right**) precipitation; (**bottom left**) wheat price; and (**bottom right**) CPI during the wet periods. The wet periods refer to the time with above-average precipitation over our study period. Red lines represent the trends, and the green envelopes provide the 95% confidence interval areas.

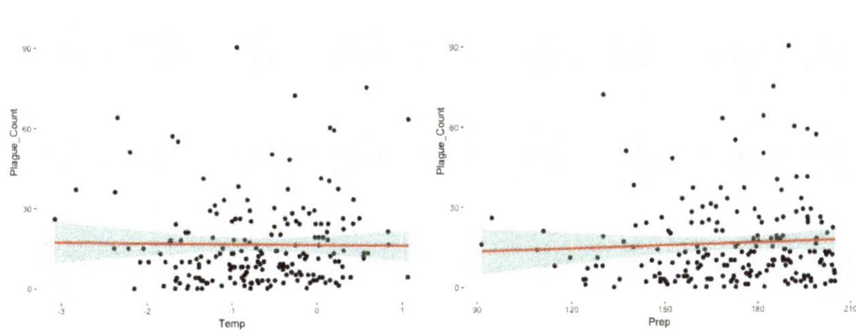

Figure 7. *Cont.*

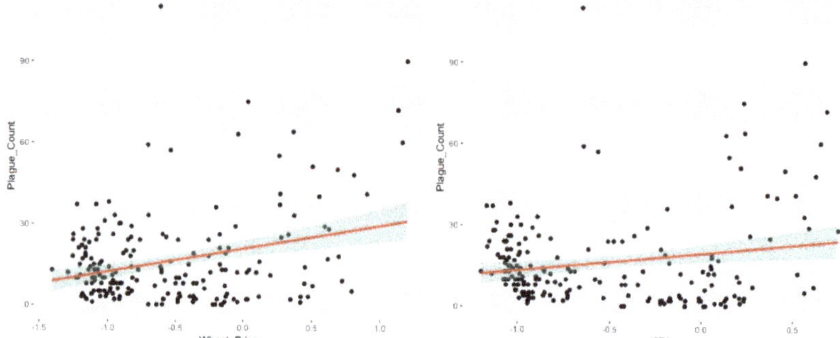

Figure 7. Long-term trend of plague outbreaks with (**top left**) temperature anomaly; (**top right**) precipitation; (**bottom left**) wheat price; and (**bottom right**) CPI during the dry periods. The dry periods refer to the time with below-average precipitation over our study period. Red lines represent the trends, and the green envelopes provide the 95% confidence interval areas.

4. Discussion

Given that climate change and the risk of plague outbreak are closely coupled [28], actions need to focus on the pathways and patterns of climate-plague nexus. In particular, the mechanism of those climatic effects—whether they are direct and/or indirect, remain largely unclear and understudied so far. In this study, we found that the effect of climatic forcing on the temporal distribution of plague outbreak was solely indirect in nature. Furthermore, the long-term trend of climate-plague nexus was only significant in the cold and wet periods.

4.1. Direct and Indirect Paths Embedded in the Climate-Plague Nexus

The statistical results from SEM led us to similar conclusions: (1) that climate has an indirect influence on plague dynamics; (2) that economic factors have a direct influence on plague dynamics; (3) that climate has a direct influence on economic factors, also meaning that the influence of climate on plague dynamics is mediated by economic systems. More specifically, the effect of temperature on plague dynamics is indirect but as a whole significant, whilst the effect of precipitation on plague dynamics is indirect but considerably insignificant as a whole.

Although many disease outbreaks are characteristically associated with climate directly [29,30], the indirect pathway is not uncommon [31]. However, the recognition of such an indirect climatic effect has not received significant attention in academia. Most often, previous studies considered climatic influence as a direct indicator to plague activities without testing the potential existence of an indirect pathway. Our results challenged the previous perspective and demonstrated that the economic system is an important element of the indirect effect of climate on plague outbreak. In fact, our SEM approach could not detect any direct climatic effect on plague activity; instead, only temperature displayed total effect on plague dynamics, and the pathways of temperature influence and precipitation influence were all indirect. Indeed, previous studies suggested that high summer temperature could inactivate human plague occurrence in the case of the United States [32,33] and may cause the reduction of flea survival, early-stage development, reproduction rate, and the ability to transmit the disease [34,35]. The influence of precipitation was previously well documented but defined in a complicated manner. The trend of precipitation was depicted as a positive and linear indicator of plague outbreak in the U.S. [36]. Nonetheless, the correlation is negative in Vietnam and Uganda, where dry seasons favored the risk of plague outbreak to a greater extent than wet seasons [37,38]. However, regardless of the disparity in the study area, the variation in climate (both temperature and precipitation) did not post any direct effect on plague dynamics in our study when we held the economic parameters constant.

Academically, much more is known about how climate affects economic systems, or how economic attributes led to epidemics–partly because climate-epidemics study is a comparatively new subject for researchers. A series of work by Pei et al. [16,39–41] is most prominent in justifying the casual pathway from cooling to the shrinkage of crop productivity, and thus, shock in the agricultural market and the stability of agrarian society. In fact, before the time of the Industrial Revolution, when most people were farmers, it is understandable that an agrarian society relies heavily on "good climate". The study by Zhang et al. [12] extended further the idea of climate-induced food shortage to malnutrition and hypothesized a declined cohort immunity through examining the time series of human height in Europe. Moreover, because of malnutrition, the undernourished population would easily be more susceptible to infectious diseases through dysfunctional immunological responses [42]. Thus, it is possible, with the support of past literatures, to demonstrate that climatic variables could be hypothetically correlated to increasing plague activities through their commands on the agro-economy. However, from our SEM result, only temperature and precipitation variations were deemed significant in such a pathway, with PDSI coming up short in showing a statistically significant relationship with plague dynamics both directly and indirectly. As a result, since not all the climatic variables we tested in this analysis were found to be correlated, additional climatic variables, or the term "climatic influence" should be considered carefully when we describe their impact on plague outbreak.

Our study also delineated the indirect effect of precipitation on plague outbreak in time series. In fact, previous results once questioned the role of precipitation in producing any significant change in plague dynamics [2]. Yet, those studies might have only considered the total effect of precipitation but not the indirect effect of it. Our study also highlighted the indirect positive effect of wetness on plague dynamics. However, our SEM results indicate that this association could be weak, yet significant, in magnitude. The reason behind this might lie in the contextual variations in the agricultural markets or major crop types. In this study, we used wheat price as a representation of crop market and Europe as the study area, and it could be that wheat is more sensitive to wetness than dryness and historical Europe was capable of absorbing shocks in mild climatic variations, and thus, generating the result we saw from our analysis. In an alternative study adopting the same methodology of ours but working on a region with a completely different market and key crop types, the role of precipitation in influencing plague dynamics might also appear different.

Furthermore, our results, suggest that the sustainability and resilience of societies to climate change matters for their survival against the plague. Some researchers identify climate as the direct driver of plague activities in European history, with such climate-plague tropes supported by robust palaeoclimatic and reconstructed disease data. Yet, those hypotheses might have simplified social responses and failed to account for the complexity of disease dynamics in human society. To a certain extent, the lack of synchronous climate-driven plague outbreak mechanisms in China in the work of Xu et al. [4] should have already implied profound regional variations in plague resilience and buffers to climatic variability. Such recognition of regional variation in plague resilience had been stifled by treating the disease dynamics of plague to climatic variation as a homogenous and universal entity. Statistical analyses that assumed a single, homogenous and direct response of plague dynamics to climate forcing were therefore at odds with empirical data relevant only to a certain spatial level so that the results might appear dichotomous, inconsistent, mild, or insignificant. Consequently, there is a dissonance between method and hypothesis.

Five important caveats should be noted in interpreting our study results. First, our study results did not refute the role of climatic variation in any previous studies, although our study showed that climatic variation was not directly related to plague dynamics. Our intent was to demonstrate that the effects of climate on the economic system exert a stronger influence on temporal plague outbreak patterns, and to call for addressing the role of social responses to plague outbreak. Second, the disease dynamics of plague remained extremely complex. Our study is pilot in nature to outline the plausible indirect effect of climate on the plague outbreak dynamics at our study scale. Yet, the interpretation of such results should not be overgeneralized to plague dynamics in other contexts such as virus-host relationship, rodent plague outbreak dynamics, host-to-human transmission, and so on. Third, the indirect effect of climate change on plague dynamics might not only bypass social response. For example, Yue, et al. [43] explored the influence of trade routes on plague spreading patterns. Despite the fact that they assumed the transportation route as static over their study period, one can easily forecast the evolution of transportation could also contribute to plague dynamics in time. Forth, in a similar sense, the effect of other climatic variables on the plague dynamics should await further analysis. Plague dynamics might not only be influenced by temperature and precipitation. Instead, recent progress in climate-plague nexus had frequently suggested the role of large-scale climatic phenomenon [44,45]. Finally, this paper focused mainly on the climate-plague nexus at the macro-scale. However, we should not overlook the importance of studying micro-regions in Europe during this outbreak. A series of recent work [46–50] has called for attention between the balanced view of macro-scale study of plague dynamics and micro-perspective of it in medieval Europe.

4.2. Long-Term Trends of the Climate-Plague Nexus

Long-term trends between different variables suggested that human plague dynamics is sensitive to particular climatic situations, and economic attributes are potentially more consistent in influencing plague dynamics. To be more specific, the long-term trend of human plague activities with different predictors suggested that only temperature remained significant in correlating with plague outbreak patterns in cold and wet climate, whilst the effect of precipitation showed no significant relationship with plague directly in any sensitivity tests. Furthermore, the effect of temperature displayed no total or indirect relationship with plague outbreak when the climate is warm and dry. Similarly, the increasing temperature only promoted the drop of wheat price during cold and wet periods but made no significant impact on the wheat market when the climate was already warm and dry. In comparison, decreasing precipitation also contributed to the lowering of wheat price in cold and wet periods. On the one hand, the above long-term dynamics are consistent with and supportive of our SEM results. On the other hand, the pattern might also indicate that the direct/indirect/total effects of climate variation were spatio-temporally selective and variable-dependent. Thus, such reshaping of the idea of direct/indirect climate-plague nexus at the temporal domain might also be applicable in similar fashion at the spatial domain, or in a multi-scalar study [51]. Second, the results suggested that cooling during cold times and growing wetness in wet times could stimulate plague outbreak. In a similar vein, extreme coldness and flooding were the contributors to plague outbreak in Europe [2]. Third, to expand on the previous point, our result might have explained the insignificant climate-plague association in some other studies. To a large extent, the meticulous picture of the climate-plague association we obtained from this study was attributable to the centennial-scale historical panel data applied. Other studies that utilize shorter time frames or contemporary plague outbreak might unavoidably fall onto the "warm period", which provided a null result, rejecting the legitimacy of climate-plague relationship.

In addition, both wheat price and CPI exhibited a positive relationship with human plague outbreak regardless of any climatic situations. The consistent relationship between economic parameters and plague might seem persuasive in confirming their dominant role in governing plague dynamics. Yet, one should also consider that plague dynamics, or any other pandemics, would certainly shock agricultural markets and living standards [52]. Therefore, a cyclic correlation between epidemics and economic systems might reinforce each other, making their correlation become consistent and stable. Besides, factors such as wars, social turmoils, urbanizations, and famine could have a fundamental impact on both agricultural markets and infectious disease outbreak [10]. Based simply on the analysis performed in this study, it remained unknown whether socio-economic variables were more important than climatic variables in driving plague dynamics, and that whether factors like wars and urbanization might play a part in climate-plague nexus. Yet, given that our analytic results could address the consistent significant correlation between economic changes and plague dynamics, further research for the aforementioned questions should be warranted. As our analysis focused on the long-term relationship of climate-plague nexus over a continental spatial unit, future study could be arranged at case studies in the modern context with a smaller spatial study unit to capture the potential variation of climate-plague nexus with a moving scalar window.

5. Conclusions

The identification of the direct and indirect relationship between climate change and human plague dynamics is a topic of high interest in the context of surging researches over the potential elevating risk of plague outbreak under climate change. Although many studies have revealed the possible linkages in the climate-plague nexus and have highlighted the concern of scale-dependent variability in climate-driven plague dynamics in the spatio-temporal dimension [45], they seldom explicitly consider the possibility that the relationship is an indirect one. In this study, we applied SEM to quantitatively justify that the casual pathway from climate change to plague dynamics in historical Europe was not a direct one but was mediated by climate-driven economic changes. In a nutshell, the influence of temperature was only significant in the cool and wet periods, corresponding to the total effect and indirect effect of temperature in SEM; whilst the influence of precipitation on human plague dynamics seemed to be indirect in nature and was significant only in cold climate. The study evidenced that climate-driven economic changes, rather than climate change alone, were the direct cause of human plague outbreak. The investigation on such indirect influence of climate change on human plague dynamics should receive more attention in the future.

Author Contributions: R.P.H.Y. and H.F.L. designed research; R.P.H.Y. and H.F.L. performed research; R.P.H.Y. and H.F.L. analyzed data; and R.P.H.Y. and H.F.L. wrote the paper. Both authors read and approved the final manuscript. All authors have read and agreed to the published version of the manuscript.

Funding: This study is supported by the Improvement on Competitiveness in Hiring New Faculties Funding Scheme (4930900) and Direct Grant for Research 2018/19 (4052199) of the Chinese University of Hong Kong.

Conflicts of Interest: The authors declare no conflict of interest.

Appendix

Table A1. Linear regression results of the long-term relationship between different sets of variable employed in the path analysis, AD1347–1760 (n = 414).

Variable A	Variable B	Coef	F	R^2
Temperature anomaly	Plague outbreak	−0.007 ***	8.24	0.0196
Precipitation	Plagueoutbreak	−0.090	0.60	0.0014
Wheat price	Plague outbreak	0.0126 ***	60.57	0.1282
CPI	Plague outbreak	0.0062 ***	16.20	0.0378
Temperature anomaly	Wheat price	−0.319 ***	23.81	0.0546
Precipitation	Wheat price	10.519 ***	10.29	0.0244
Wheat price	CPI	1.029 ***	2597.16	0.8631

Significance level: *** $p < 0.001$.

Table A2. Linear regression results of the long-term relationship between different sets of variable used in path analysis during the cool periods of AD1347–1760 (n = 319).

Variable A	Variable B	Coef	F	R^2
Temperature anomaly	Plague outbreak	−0.007 ***	11.73	0.0357
Precipitation	Plague outbreak	−0.090	0.08	0.0002
Wheat price	Plague outbreak	0.0125 ***	48.11	0.1318
CPI	Plague outbreak	0.0054***	9.70	0.0297
Temperature anomaly	Wheat price	−0.229 ***	14.90	0.0449
Precipitation	Wheat price	13.655***	13.84	0.0418
Wheat	CPI	1.025 ***	1875.74	0.8554

Significance level: *** $p < 0.001$.

Table A3. Linear regression results of the long-term relationship between different sets of variable used in path analysis during the warm periods of AD1347–1760 (n = 95).

Variable A	Variable B	Coef	F	R^2
Temperature anomaly	Plague outbreak	−0.0007	0.08	0.0009
Precipitation	Plague outbreak	2.503	0.11	0.0012
Wheat price	Plague outbreak	0.0122 ***	12.04	0.1146
CPI	Plague outbreak	0.0093 ***	8.66	0.0852
Temperature anomaly	Wheat price	0.020	0.09	0.0010
Precipitation	Wheat price	2.503	0.11	0.0012
Wheat price	CPI	1.065 ***	691.12	0.8814

Significance level: *** $p < 0.001$.

Table A4. Linear regression results of the long-term relationship between different sets of variable used in path analysis during the wet periods of AD1347–1760 (n = 217).

Variable A	Variable B	Coef	F	R^2
Temperature anomaly	Plague outbreak	−0.0119 ***	12.9	0.0566
Precipitation	Plague outbreak	0.088	0.00	0.0000
Wheat price	Plague outbreak	0.0147 ***	47.82	0.1819
CPI	Plague outbreak	0.0064 ***	8.66	0.0852
Temperature anomaly	Wheat price	−0.524 ***	32.18	0.1302
Precipitation	Wheat price	0.879	0.00	0.0000
Wheat	CPI	1.020 ***	1166.35	0.8444

Significance level: *** $p < 0.001$.

Table A5. Linear regression results of the long-term relationship between different sets of variable used in path analysis during the dry periods of AD1347–1760 (n = 197).

Variable A	Variable B	Coef	F	R^2
Temperature anomaly	Plague outbreak	−0.001	0.06	0.0566
Precipitation	Plague outbreak	3.679	1.73	0.0088
Wheat price	Plague outbreak	0.0104 ***	18.86	0.0882
CPI	Plague outbreak	0.0061 **	7.46	0.0369
Temperature anomaly	Wheat price	−0.142	2.38	0.0120
Precipitation	Wheat price	3.68	1.73	0.0088
Wheat	CPI	1.038 ***	1377.92	0.8760

Significance level: *** $p < 0.001$; ** $p < 0.01$.

References

1. Schmid, B.V.; Büntgen, U.; Easterday, W.R.; Ginzler, C.; Walløe, L.; Bramanti, B.; Stenseth, N.C. Climate-driven introduction of the Black Death and successive plague reintroductions into Europe. *Proc. Natl. Acad. Sci. USA* **2015**, *112*, 3020–3025. [CrossRef] [PubMed]
2. Yue, R.P.; Lee, H.F. Pre-industrial plague transmission is mediated by the synergistic effect of temperature and aridity index. *BMC Infect. Dis.* **2018**, *18*, 134. [CrossRef] [PubMed]
3. Stenseth, N.C.; Samia, N.I.; Viljugrein, H.; Kausrud, K.L.; Begon, M.; Davis, S.; Leirs, H.; Dubyanskiy, V.; Esper, J.; Ageyev, V.S. Plague dynamics are driven by climate variation. *Proc. Natl. Acad. Sci. USA* **2006**, *103*, 13110–13115. [CrossRef] [PubMed]
4. Xu, L.; Liu, Q.; Stige, L.C.; Ari, T.B.; Fang, X.; Chan, K.S.; Wang, S.; Stenseth, N.C.; Zhang, Z. Nonlinear effect of climate on plague during the third pandemic in China. *Proc. Natl. Acad. Sci. USA* **2011**, *108*, 10214–10219. [CrossRef]
5. Xu, L.; Stige, L.C.; Kausrud, K.L.; Ben Ari, T.; Wang, S.; Fang, X.; Schmid, B.V.; Liu, Q.; Stenseth, N.C.; Zhang, Z. Wet climate and transportation routes accelerate spread of human plague. *Proc. R. Soc. B Biol. Sci.* **2014**, *281*, 20133159. [CrossRef]

6. Stapp, P.; Antolin, M.F.; Ball, M. Patterns of extinction in prairie dog metapopulations: Plague outbreaks follow El Ninì events. *Front. Ecol. Environ.* **2004**, *2*, 235–240.

7. Kreppel, K.S.; Caminade, C.; Telfer, S.; Rajerison, M.; Rahalison, L.; Morse, A.; Baylis, M. A non-stationary relationship between global climate phenomena and human plague incidence in Madagascar. *PLoS Negl. Trop. Dis.* **2014**, *8*, e3155. [CrossRef]

8. Ben Ari, T.; Gershunov, A.; Gage, K.L.; Snäll, T.; Ettestad, P.; Kausrud, K.L.; Stenseth, N.C. Human plague in the USA: The importance of regional and local climate. *Biol. Lett.* **2008**, *4*, 737–740. [CrossRef]

9. Zhang, Z.; Li, Z.; Tao, Y.; Chen, M.; Wen, X.; Xu, L.; Tian, H.; Stenseth, N.C. Relationship between increase rate of human plague in China and global climate index as revealed by crosss-pectral and cross-wavelet analyses. *Integr. Zool.* **2007**, *2*, 144–153. [CrossRef]

10. Zhang, D.D.; Brecke, P.; Lee, H.F.; He, Y.Q.; Zhang, J. Global climate change, war, and population decline in recent human history. *Proc. Natl. Acad. Sci. USA* **2007**, *104*, 19214–19219. [CrossRef]

11. Zhang, D.D.; Lee, H.F.; Wang, C.; Li, B.; Zhang, J.; Pei, Q.; Chen, J. Climate change and large-scale human population collapses in the pre-industrial era. *Glob. Ecol. Biogeogr.* **2011**, *20*, 520–531. [CrossRef]

12. Zhang, D.D.; Lee, H.F.; Wang, C.; Li, B.; Pei, Q.; Zhang, J.; An, Y. The causality analysis of climate change and large-scale human crisis. *Proc. Natl. Acad. Sci. USA* **2011**, *108*, 17296–17301. [CrossRef] [PubMed]

13. Lee, H.F.; Zhang, D.D. A tale of two population crises in recent Chinese history. *Clim. Chang.* **2013**, *116*, 285–308. [CrossRef]

14. Tian, H.; Yan, C.; Xu, L.; Büntgen, U.; Stenseth, N.C.; Zhang, Z. Scale-dependent climatic drivers of human epidemics in ancient China. *Proc. Natl. Acad. Sci. USA* **2017**, *114*, 12970–12975. [CrossRef] [PubMed]

15. Duncan, C.; Scott, S. The key role of nutrition in controlling human population dynamics. *Nutr. Res. Rev.* **2004**, *17*, 163–175. [CrossRef]

16. Pei, Q.; Zhang, D.D.; Li, G.; Lee, H.F. Climate change and the macroeconomic structure in pre-industrial Europe: New evidence from wavelet analysis. *PLoS ONE* **2015**, *10*, e0126480. [CrossRef]

17. Büntgen, U.; Ginzler, C.; Esper, J.; Tegel, W.; McMichael, A.J. Digitizing historical plague. *Clin. Infect. Dis.* **2012**, *55*, 1586–1588. [CrossRef]

18. Lee, H.F.; Zhang, D.D.; Pei, Q.; Jia, X.; Yue, R.P.H. Demographic impact of climate change on northwestern China in the late imperial era. *Quat. Int.* **2016**, *425*, 237–247. [CrossRef]

19. Lee, H.F. Internal wars in history: Triggered by natural disasters or socio-ecological catastrophes? *Holocene* **2018**, *28*, 1071–1081. [CrossRef]

20. Büntgen, U.; Tegel, W.; Nicolussi, K.; McCormick, M.; Frank, D.; Trouet, V.; Kaplan, J.O.; Herzig, F.; Heussner, K.-U.; Wanner, H. 2500 years of European climate variability and human susceptibility. *Science* **2011**, *331*, 578–582. [CrossRef]

21. Yue, R.P.H.; Lee, H.F. Climate change and plague in European history. *Sci. China Earth Sci.* **2018**, *61*, 163–177. [CrossRef]

22. Cook, E.R.; Seager, R.; Kushnir, Y.; Briffa, K.R.; Büntgen, U.; Frank, D.; Krusic, P.J.; Tegel, W.; van der Schrier, G.; Andreu-Hayles, L. Old World megadroughts and pluvials during the Common Era. *Sci. Adv.* **2015**, *1*, e1500561. [CrossRef] [PubMed]

23. Allen, R. Allen-Unger Database: European Commodity Prices 1260–1914. 2007. Available online: http://www2.history.ubc.ca/unger/htmfiles/newgrain.htm (accessed on 20 February 2020).

24. Shipley, B. *Cause and Correlation in Biology: A User's Guide to Path Analysis, Structural Equations and Causal Inference with R*; Cambridge University Press: Cambridge, UK, 2016.

25. Grace, J.B.; Schoolmaster, D.R., Jr.; Guntenspergen, G.R.; Little, A.M.; Mitchell, B.R.; Miller, K.M.; Schweiger, E.W. Guidelines for a graph-theoretic implementation of structural equation modeling. *Ecosphere* **2012**, *3*, 1–44. [CrossRef]

26. Jansson, R. Global patterns in endemism explained by past climatic change. *Proc. R. Soc. Lond. Ser. B Biol. Sci.* **2003**, *270*, 583–590. [CrossRef]

27. Rohlf, F.J.; Sokal, R.R. *Biometry: The Principles and Practice of Statistics in Biological Research*; Freeman: New York, NY, USA, 1981.

28. Stenseth, N.C.; Atshabar, B.B.; Begon, M.; Belmain, S.R.; Bertherat, E.; Carniel, E.; Gage, K.L.; Leirs, H.; Rahalison, L. Plague: Past, present, and future. *PLoS Med.* **2008**, *5*, e3. [CrossRef]

29. Colwell, R.R.; Patz, J.A. Climate, Infectious Disease, and Health. In *An Interdisciplinary Perspective*; American Academy of Microbiology: Washington, DC, USA, 1998.

30. Gubler, D.J.; Reiter, P.; Ebi, K.L.; Yap, W.; Nasci, R.; Patz, J.A. Climate variability and change in the United States: Potential impacts on vector-and rodent-borne diseases. *Environ. Health Perspect.* **2001**, *109*, 223–233.

31. WHO. *Using Climate to Predict Infectious Disease Epidemics*; World Health Organization: Geneva, Switzerland, 2005.

32. Enscore, R.E.; Biggerstaff, B.J.; Brown, T.L.; Fulgham, R.E.; Reynolds, P.J.; Engelthaler, D.M.; Levy, C.E.; Parmenter, R.R.; Montenieri, J.A.; Cheek, J.E. Modeling relationships between climate and the frequency of human plague cases in the southwestern United States, 1960–1997. *Am. J. Trop. Med. Hyg.* **2002**, *66*, 186–196. [CrossRef]

33. Ari, T.B.; Gershunov, A.; Tristan, R.; Cazelles, B.; Gage, K.; Stenseth, N.C. Interannual variability of human plague occurrence in the Western United States explained by tropical and North Pacific Ocean climate variability. *Am. J. Trop. Med. Hyg.* **2010**, *83*, 624–632. [CrossRef]

34. Cavanaugh, D.C. Specific effect of temperature upon transmission of the plague bacillus by the oriental rat flea, Xenopsylla cheopis. *Am. J. Trop. Med. Hyg.* **1971**, *20*, 264–273. [CrossRef]

35. Krasnov, B.; Khokhlova, I.; Fielden, L.; Burdelova, N. Development rates of two Xenopsylla flea species in relation to air temperature and humidity. *Med. Vet. Entomol.* **2001**, *15*, 249–258. [CrossRef]

36. Parmenter, R.R.; Yadav, E.P.; Parmenter, C.A.; Ettestad, P.; Gage, K.L. Incidence of plague associated with increased winter-spring precipitation in New Mexico. *Am. J. Trop. Med. Hyg.* **1999**, *61*, 814–821. [CrossRef] [PubMed]

37. Moore, S.M.; Monaghan, A.; Griffith, K.S.; Apangu, T.; Mead, P.S.; Eisen, R.J. Improvement of disease prediction and modeling through the use of meteorological ensembles: Human plague in Uganda. *PLoS ONE* **2012**, *7*, e44431. [CrossRef]

38. Pham, H.V.; Dang, D.T.; Tran Minh, N.N.; Nguyen, N.D.; Nguyen, T.V. Correlates of environmental factors and human plague: An ecological study in Vietnam. *Int. J. Epidemiol.* **2009**, *38*, 1634–1641. [CrossRef] [PubMed]

39. Pei, Q.; Zhang, D.D.; Lee, H.F.; Li, G. Climate change and macro-economic cycles in pre-industrial Europe. *PLoS ONE* **2014**, *9*, e88155. [CrossRef]

40. Pei, Q.; Zhang, D.D.; Li, G.; Winterhalder, B.; Lee, H.F. Epidemics in Ming and Qing China: Impacts of changes of climate and economic well-being. *Soc. Sci. Med.* **2015**, *136*, 73–80. [CrossRef]

41. Pei, Q.; Zhang, D.D.; Li, G.; Forêt, P.; Lee, H.F. Temperature and precipitation effects on agrarian economy in late imperial China. *Environ. Res. Lett.* **2016**, *11*, 064008. [CrossRef]

42. Keusch, G.T. The history of nutrition: Malnutrition, infection and immunity. *J. Nutr.* **2003**, *133*, 336S–340S. [CrossRef]

43. Yue, R.P.; Lee, H.F.; Wu, C.Y. Trade routes and plague transmission in pre-industrial Europe. *Sci. Rep.* **2017**, *7*, 1–10. [CrossRef]

44. Yue, R.P.; Lee, H.F. Drought-induced spatio-temporal synchrony of plague outbreak in Europe. *Sci. Total Environ.* **2020**, *698*, 134138. [CrossRef]

45. Ben-Ari, T.; Neerinckx, S.; Gage, K.L.; Kreppel, K.; Laudisoit, A.; Leirs, H.; Stenseth, N.C. Plague and climate: Scales matter. *PLoS Pathog.* **2011**, *7*, e1002160. [CrossRef]

46. Roosen, J.; Curtis, D.R. Dangers of noncritical use of historical plague data. *Emerg. Infect. Dis.* **2018**, *24*, 103. [CrossRef]

47. Roosen, J.; Curtis, D.R. The 'light touch' of the Black Death in the Southern Netherlands: An urban trick? *Econ. Hist. Rev.* **2019**, *72*, 32–56. [CrossRef] [PubMed]

48. Van Bavel, B.J.; Curtis, D.R.; Hannaford, M.J.; Moatsos, M.; Roosen, J.; Soens, T. Climate and society in long-term perspective: Opportunities and pitfalls in the use of historical datasets. *Wiley Interdiscip. Rev. Clim. Chang.* **2019**, *10*, e611.

49. Curtis, D.R. *Coping with Crisis: The Resilience and Vulnerability of Pre-Industrial Settlements*; Ashgate Publishing, Ltd.: Farnham, UK, 2014.

50. Van Bavel, B.; Curtis, D. Better Understanding Disasters by Better Using History: Systematically Using the Historical Record as One Way to Advance Research into Disasters. *Int. J. Mass Emergencies Disasters* **2016**, *34*, 143–169.

51. Lee, H.F.; Fei, J.; Chan, C.Y.; Pei, Q.; Jia, X.; Yue, R.P. Climate change and epidemics in Chinese history: A multi-scalar analysis. *Soc. Sci. Med.* **2017**, *174*, 53–63. [CrossRef] [PubMed]
52. Voigtländer, N.; Voth, H.J. The three horsemen of riches: Plague, war, and urbanization in early modern Europe. *Rev. Econ. Stud.* **2013**, *80*, 774–811. [CrossRef]

Article

The Application of a Decision Tree and Stochastic Forest Model in Summer Precipitation Prediction in Chongqing

Bo Xiang [1], Chunfen Zeng [2,3,*], Xinning Dong [1] and Jiayue Wang [2]

[1]   Chongqing Climate Center, Chongqing 401147, China; actnowx@126.com (B.X.); dxn98mry@sohu.com (X.D.)
[2]   School of Geography and Tourism, Chongqing Normal University, Chongqing 401331, China; jiayuewyy@163.com
[3]   Key Laboratory of Surface Processes and Environment Remote Sensing in the Three Gorges Reservoir Chongqing Normal University, Chongqing 401331, China
*   Correspondence: cfzeng18@163.com

Received: 12 March 2020; Accepted: 10 May 2020; Published: 14 May 2020

Abstract: Meteorological disasters are the result of the interaction of multiple factors and multiple systems. In order to improve the accuracy of prediction, it is necessary not only to consider the characteristics and cycles of each subsystem, but also to study the interaction of all systems. Based on the summer precipitation data and 130 circulation indexes of 34 national meteorological observation stations in Chongqing from 1961 to 2010, the prediction model of Chongqing summer precipitation was established based on the decision tree and the stochastic forest algorithm based on machine learning, and the prediction test of 2011–2018 was carried out independently by the model. Compared with the results of the single-factor prediction model, the trend consistency rate increased by 37.5% and 12.5% respectively. In addition, when using the random forest model to predict summer precipitation in Chongqing from 2014 to 2018, the 5-year average Ps, Cc and PC scores were 84.6, 0.27 and 67.1, respectively, which were significantly improved compared with 72.4, −0.12 and 52.9 of the current climate forecasting methods, and the forecast quality of the random forest was relatively stable. The multi-system collaborative impact model based on decision tree and random forest algorithm can achieve high accuracy and stability. Thus, this method can not only be an effective means for the diagnosis and prediction of climate causes, but also has a good theoretical and practical value for the prediction of extreme disasters.

Keywords: decision tree; random forest; precipitation prediction; machine learning

1. Introduction

Since summer precipitation is of great concern regarding meteorological disasters, a lot of research work has been done on the influence of climate system subsystem changes and their interactions on summer precipitation. Some progress has been made in the study of the characteristics, causes and prediction methods of summer precipitation in Chongqing and its surrounding areas. Li [1–3] analyzed the characteristics of summer precipitation, drought and flood in the eastern part of southwest China, and pointed out that it had obvious inter-annual and inter-decadal changes. Zhou et al. [4] studied the basic climatic characteristics of summer precipitation in the three gorges reservoir area, and the results showed that the summer precipitation in the three gorges reservoir area had a good consistency, the frequency of drought years was significantly higher than that of flood years, and the summer precipitation in the three gorges reservoir area had an obvious inter-decadal variation. Ma Zhenfeng [5] analyzed the main physical factors affecting summer precipitation in southwest China, such as plateau factors, westerly belt system, subtropical high and other factors, and established a summer precipitation prediction model with a certain physical basis on this basis, which achieved good results in precipitation prediction in flood season in recent years. Zhang Qiang et al. [6] analyzed the correlation between Sea Surface Temperature (SST) index and drought and flood disaster in the upper reaches of the Yangtze river, showing that the occurrence of El Nino event increases the probability of drought in the upper reaches of the Yangtze river, while the occurrence of La Nina event increases the probability of waterlogging in the upper reaches of the Yangtze river. Liu De et al. [7] analyzed the characteristics of Eurasian circulation in summer rainfall in Chongqing, and established a conceptual model for forecasting summer precipitation in Chongqing by using circulation index in key areas in early winter.

In recent years, artificial intelligence technology has also begun to be applied in the field of atmospheric science such as severe convection weather forecast and climate prediction. There are many applications of machine learning in severe convection weather prediction. In 2017, Shenzhen Meteorological Bureau and Alibaba jointly organized the CIKM data science competition themed "smart city, smart country", and made climate precipitation forecast with radar images. Xiu Yuanyuan et al. [8] used machine learning supervised learning model support vector machine SVM to identify and forecast severe convection weather. Sun Quande et al. [9] showed the potential of machine learning methods in improving local accurate weather prediction. Li Wenjuan et al. [10] founded that the physical significance of factors selected by the random forest algorithm was relatively clear. In the field of climate, researchers have used artificial intelligence systems to help them rank climate models over the past few years [11] to detect hurricanes and other extreme weather events in real and simulated climate data, and thus find new climate models. The above study is based on considering the influence of a single system or physical factor on precipitation in and around Chongqing. The effects of anomalies of multiple systems or physical factors on precipitation in Chongqing are considered. In fact, due to the non-linear and chaotic nature of the climate system, the factors that affect the precipitation prediction constitute the comprehensive effects of many sea temperatures (ENSO, Kuroshio, etc.), plateau snow, land surface temperature, volcanic activity, astronomical factors, monsoon, subtropical high, high resistance and plateau topography. We aim to analyze the synergistic effect of the factors that lead to the precipitation change through sorting, statistics, analysis and processing of big data, machine learning, etc., and distinguish which factors are excellent forecasting factors, and the weight of these excellent factors in different regions, that is, how much forecasting information these factors can provide. If these issues are resolved, precipitation prediction will become possible and credible.

As an excellent representative of the machine learning algorithm, the decision tree model adopts recursive segmentation technology to continuously divide the data space into different subsets so as to detect the potential structure, important patterns and relationships of data [12]. Compared with traditional parametric statistical methods, the decision tree model does not need to make assumptions about the relationship between independent variables and dependent variables in advance, and it can effectively overcome the multi-collinearity of independent variables. However, the results of a single decision tree are unstable and prone to overfitting. The random forest model builds decision trees by randomly extracting some samples from the original samples through Bootstrap sampling technology, and combines multiple decision trees to effectively avoid overfitting [13]. At present, decision trees and random forest algorithms are more and more widely used in meteorology. Shi Dawei et al. [14] used the decision tree algorithm to establish a more accurate classification and prediction model for road icing disaster. Shi Yimin et al. [15] studied the classification and prediction model of regional summer precipitation days based on the data mining Classification and Regression Tree (CART) algorithm. Qin Pengcheng et al. [16] Hubei rapeseed yield limiting factor analysis based on a decision tree and random forest model also achieved good application.

Based on the actual forecast business, this paper adopts the decision tree classification method to establish the precipitation prediction model with multi-factor collaborative influence for the average summer precipitation in Chongqing. Based on decision tree modeling, random forest is used to conduct an integrated prediction test, and evaluate its prediction effects.

2. Materials and Methods

2.1. Materials

The meteorological data used in this paper were obtained through the Meteorological Unified Service Interface Community of China Meteorological Administration. The decision tree method uses the regional average precipitation of 34 national meteorological observing stations in Chongqing (Figure 1), while random forest analyzes the precipitation of 34 national meteorological observing stations.

Figure 1. Map of 34 national meteorological stations in Chongqing.

The circulation index in this article comes from the business meteorological network [17–20], respectively, including atmospheric circulation index 88 such as the subtropical high, east Asian trough, polar vortex, Eurasian circulation type, characteristics, Pacific Ocean trade winds etc., SST index 26 such as ENSO (various districts and types), warm pools, Indian Ocean, Tide area, Kuroshio area et al., cold air and typhoon, Southern Oscillation Index (SOI), The Pacific Decadal Oscillation (PDO), Quasi-biennial Oscillation (QBO), subsurface SST index and another 16; in total, 130 items. Among them, the SST index in this article includes 26 SST indexes and 16 other indexes including multivariable ENSO index, decadal oscillation in the north Pacific, meridional mode SST in the Atlantic Ocean, quasi-biennial oscillation, EU300T_130E, EU300T_160E, EU300T_180W, and Atlantic SST triples. In this article, summer refers to June to August.

2.2. Methods

The machine learning algorithms adopted in this article include decision tree [14,16,21], and random forest [22–24]. The evaluation methods are Prediction Consistency (Pc), Prediction Score (Ps) and Correlation Coefficient (Cc) [25–27], which are used in the China Meteorological Administration (CMA). The modeling period is from 1961 to 2010, and the independent inspection and evaluation period is from 2011 to 2018. The decision tree and stochastic forest model are multi-factor prediction models using 130 circulation indexes, while the prediction model established by using any one of the 130 circulation indexes is a single-factor prediction model.

2.2.1. The Decision Tree

Yang et al. [28] introduced the basic concepts and common algorithms of decision tree, which can be used to form classification and predictive model. Suppose $D = \{(x_1, y_1), (x_2, y_2), \ldots, (x_n, y_n)\}$, including $x_i = \left(x_i^{(1)}, x_i^{(2)}, \ldots, x_i^{(n)}\right)^T$ as input variables (circulation index), n is the number of features (the summer model is 130, the winter model is 34), $y_i \in \{1, 2, \ldots, K\}$ is the category-type response variable (that is, the amount of precipitation), $i = 1, 2, \ldots, N$, N is the sample size (from 1961–2018, 58 years). Among them, 1961–2010 is the training data set and the completion of model training; 2011–2018 is the independent test data set and the independent inspection and evaluation. The goal of decision tree learning is to build a decision tree model based on a given training set to enable it to correctly classify instances. In this paper, the C4.5 algorithm of Quinlan is adopted for decision tree generation [29].

2.2.2. Random Forest

Random forest is a multifunctional machine learning algorithm. It was first proposed by Breiman, a professor of statistics at the University of California, Berkeley, in 2001, and can perform regression and classification calculations. The basic composition of the random forest is classification and regression tree (classification and regression tree) invented by Breiman and other inventions. Compared with machine learning algorithms such as neural networks, this algorithm of repeated classification and regression of binary data effectively reduces the amount of calculation. Random forest is the combination and re-aggregation of these classification trees. Random forest improves the estimation accuracy without significant increase in the calculation amount, and it is insensitive to missing values and multivariate collinearity, and can estimate up to thousands of explanatory variables, which why it is known as one of the best algorithms at present [30,31].

Random forests use the Bagging method to combine decision trees, and they use the Bootstrap sampling methods (Bootstrap method) to extract N samples from the original sample to model the decision tree. Under normal circumstances, random forests will randomly generate hundreds to thousands of decision trees. Each tree in the forest is independent, and then the most repetitive tree is selected as the final result. Since there is no need to consider constraints such as variable distribution conditions, interactions, non-linear effects and even missing values, the structure of the random forest is complex, but it is robust and easy to use [32–34].

The specific construction process of the random forest is as follows:

(1) If the size of the training set is N (50 in this paper, that is, 1961–2010), for each tree, N training samples are randomly and recursively extracted from the training set (this sampling method is called bootstrap sample method) as the training set of the tree;

(2) If the feature dimension of each sample is M, specify a constant m << M, randomly select m (20 in this paper) feature subsets from M features, and select the optimal one from these m features every time the tree splits.

(3) Every tree grows as fast as possible, and there is no pruning process.

(4) Established a large number of decision trees according to steps (1)–(3), thus forming a random forest. The classification result depends on the number of votes of the tree classifier.

In the process of building a random forest, there are two parameters that need to be set by the user according to the specific situation. In most cases, the default parameters of the model can obtain the optimal simulation results without adjustment. The term "random" in random forest refers to the two random parameters here. The introduction of these two randomness factors is crucial to the classification performance of a random forest. Due to their introduction, the random forest is not easy to fall into overfitting and has a good anti-noise ability (for example, insensitive to the default value). Therefore, the random forest models established in this paper to estimate precipitation all use default parameters.

2.2.3. Test Method

In order to test the climate prediction quality, CMA used a prediction grading score in 2010, then Ps and Cc in 2013. In order to be consistent with the current climate operations, Pc, Ps and Cc are used in this paper to test the prediction quality of the summer precipitation in Chongqing.

(1) Pc is evaluated station by station on the basis of whether the predicted and actual anomaly coincidence was consistent. The consistency rate formula is defined as follows:

$$Pc = \frac{N0}{N} \times 100\%$$

where N0 is the number of the stations with correct climate trend prediction; N is the number of stations actually participating in the assessment.

(2) P_s test method is a method that sets different weights to comprehensively test the results of climate trend prediction and anomaly level prediction. Its test score is relatively intuitive. On the basis of the correct score of trend prediction, the correct score of abnormal prediction can still be obtained, which is equivalent to giving encouragement to the abnormal forecast, and its prediction score can relatively reflect the ability and level of climate prediction.

Trend prediction is the prediction of anomaly/anomaly percentage sign. When the prediction is identical to the actual sign (0 for positive), the trend prediction is correct. Anomaly level prediction refers to the prediction that the percentage of precipitation anomaly exceeds (including) ±20% and the temperature anomaly exceeds (including) ±1 °C.

Calculation formula of Ps test method:

$$Ps = \frac{a \times N_0 + b \times N_1 + c \times N_2}{(N - N_0) + a \times N_0 + b \times N_1 + c \times N_2 + M} \times 100$$

where, N_0 is the number of stations with correct climate trend prediction; N_1 is the number of stations with correct first-order anomaly prediction; N_2 is the number of stations with correct second-order anomaly prediction; N is the actual number of participating evaluation stations; M is the number of stations where there are no secondary anomalies and the precipitation anomaly percentage $\geq 100\%$ or equal to -100% and the temperature anomaly ≥ 3 °C or ≤ -3 °C; a is the weight coefficients of climate trend terms, b is the first-order abnormal terms and c is the second-order abnormal terms. In this method, $a = 1$, $b = 2$ and $c = 4$.

(3) Cc tests the correlation of climate trend prediction products, which characterizes the degree of correlation between the forecast and the live field. The size of the correlation coefficient can indicate the correspondence between the high and low center of the forecast field and the live field. It reflects the accuracy of the prediction result and the quality of the prediction method to a certain extent. It is one of the internationally popular prediction evaluation methods. Prediction inspection and evaluation of precipitation and the temperature mainly use precipitation anomaly percentage and average temperature anomaly to calculate their correlation coefficients.

Specific calculation method:

$$Cc = \frac{\sum\limits_{i=1}^{N} (\Delta R_{fi} - \overline{\Delta R_f})(\Delta R_{0i} - \overline{\Delta R_0})}{\sqrt{\sum\limits_{i=1}^{N} (\Delta R_{fi} - \overline{\Delta R_f})^2 \sum\limits_{i=1}^{N} (\Delta R_{0i} - \overline{\Delta R_0})^2}}$$

where ΔR_{fi} is the forecast value of precipitation anomaly percentage of each station; $\overline{\Delta R_f}$ is the average value of the precipitation anomaly percentage of all stations in the region; ΔR_{0i} is the observed actual value of the precipitation anomaly percentage of all stations in the region; $\overline{\Delta R_0}$ is the average value of the observed values of precipitation anomaly percentage of all stations in the region; N is the total number of stations actually participating in the assessment.

The forecast released in this article refers to the forecast submitted by the Chongqing Climate Center to the National Climate Center to participate in the assessment of forecast quality.

3. Results and Analysis of Precipitation Prediction in Summer Test

This section may be divided by subheadings. It should provide a concise and precise description of the experimental results, their interpretation as well as the experimental conclusions that can be drawn. In the actual climate prediction of Chongqing, due to the complex and changeable terrain, it is necessary to make trend judgment on the precipitation and average precipitation of 34 stations in order to obtain the forecast data and obtain the detailed spatial distribution. The authors used decision trees and random forests for correlation analysis based on both averages and individual site data from 34 sites. The single site results of the decision tree is relatively complex, and have no significant characteristics, while the single site results of random forest analysis are relatively good. However, due to the limitation of sample number, while based on average date of 34 sites, the results of random forest are not as good as those of the decision tree method. Therefore, in this paper, the decision tree model takes the average precipitation of 34 sites as the modeling object and focuses on the collaborative influence of multiple factors. In the random forest model, 34 sites were modeled, and spatial distribution characteristics were focused. The two methods complement each other.

3.1. Decision Tree Model Test

Considering the physical factors in the summer period, IBM SPSS Modeler 18.0 was used, and the CART algorithm (the same as below) was used to model (Figure 2). It can be seen from the model that the circulation index that has a large impact on summer precipitation in Chongqing includes the western Pacific subtropical high ridge line, landfall typhoon, SST in tidal zone, the northern boundary of the North African Atlantic North American High, polar vorticity in the Atlantic European region, Indian sub-high area, and 30 hPa zonal wind.

The combination of the summer rainfall trend in Chongqing and the concurrent circulation index model based on the CART algorithm is shown in Table 1. Factors with less precipitation include factors 1–4, factors with more precipitation include factors 5–7. "+" and "−" respectively represent the positive and negative anomalies of the exponent in the condition, and the percentile in brackets is the probability of less (or more).

Using the same period index from 2011 to 2018 to predict the amount of summer precipitation in Chongqing and compare it with the observations, the results are shown in Table 2.

If the prediction only considers single-factor effects, the northern (southern) ridge of the western Pacific subtropical high (referred to as the Western Pacific subtropical high) generally corresponds to less (more) summer precipitation in Chongqing. Based on this prediction, the northern ridge of the western Pacific subtropical high in 2011, 2012, 2015 and 2018 corresponds to less precipitation, and the results in 2015 are inconsistent. In 2013, 2014, 2016 and 2017, the southward ridge of the western Pacific subtropical high corresponds to more precipitation, but only in 2014 and 2017. The total prediction accuracy was 62.5% (5/8).

When multi-factor synergy is considered, even if the western Pacific subtropical high is southerly, there may be less precipitation, as shown in case (3). In the actual prediction, 2011 and 2012 are completely in line with the situation (1). The percentages of precipitation anomalies are −30.5% and −22.1%, which are significantly less. The percentage of precipitation anomaly in 2013 was −26.1%, and the result was consistent with situation (3). If only the first two conditions of situation (3) are met, the probability of less precipitation is only 50%. In 2013, the 30 hPa zonal wind was significantly larger, which increased the probability of less precipitation to 100%. Similarly, in the collaborated multi-factor prediction, either the ridge line of the western Pacific subtropical high is northerly or southerly, there may be more precipitation, as shown in situation (5) and situation (6). The circulation index in 2014 is consistent with the result of situation (6). The probability of more precipitation is 100%, and the actual precipitation anomalies percentage is 6.3%, which is more normal. The circulation index in 2015 is consistent with the result of situation (5). The probability of excessive precipitation is 100%, and the percentage of actual precipitation anomaly is 11.7%. The circulation index in 2016 is consistent with the situation (3), which predicts less precipitation, but the actual situation is 9.5% more precipitation. 2016 is a typical El Nino year, and the anomaly of the atmospheric system caused by the SSP anomaly in the Pacific Ocean may be the possible reason for the failure of the prediction model in 2016 [30,31].

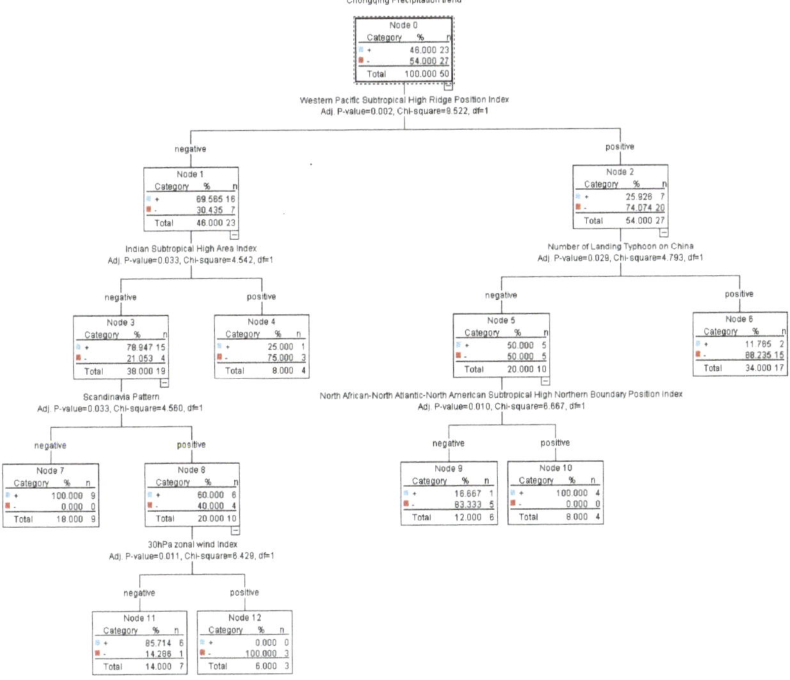

Figure 2. An analytic diagram of the relationship between the precipitation trend and circulation index in summer in Chongqing based on the CART algorithm. The '% 'represents the probability of more or less. The 'n' represents more or less annual scores (as below).

Table 1. Combination of summer precipitation trend and circulation index model based on the CART algorithm in Chongqing.

Situation	Condition 1	Condition 2	Condition 3	Condition 4
1	Northerly "+" of the western Pacific subtropical ridge (74%)	More landfall typhoons "+" (88%)		
2	Northerly "+" of the western Pacific subtropical ridge (74%)	Fewer landfall typhoons "−" (50%)	North African Atlantic north American subtropical high north boundary south (83%)	
3	Southerly "−" of the western Pacific	India's subtropical high is larger (75%)		
4	Southerly "−" of the western Pacific subtropical ridge (30%)	Sub-high area of India is relatively small "−" (21%)	Scandinavian tele-related positive "+" (40%)	30hpa zonal wind "+" (100%)
5	Northerly "+" of the western Pacific subtropical ridge (26%)	Fewer landfall typhoons "−" (50%)	North Atlantic north American subtropical high northern boundary "+" (100%)	
6	Southerly "−" of the western Pacific subtropical high ridge (70%)	India's sub-high area is smaller (79%)	Scandinavian telecorrelation negative "−" (100%)	
7	Southerly "−" of the western Pacific subtropical high ridge (70%)	India's sub-high area is smaller (79%)	Scandinavian telecorrelation positive "+" (60%)	30hPa latitudinal wind slant small "−" (86%)

Table 2. Different circulation index anomaly, precipitation prediction and observation from 2011 to 2018.

Years	Ridge Line of Western Pacific Subtropical High	Landfall Typhoon	India Sub-High Area	North Atlantic North American Subtropical High Northern Border	Scandinavian Telepathic Type	30 hpa Zonal Wind	Prediction	Observation (%)
2011	2.22	0.14	−0.04	0.48	0.04	−0.33	less	−30.5
2012	2.07	0.18	0.07	0.91	−0.2	−5.69	less	−22.1
2013	−0.10	0.81	0.09	1.5	0	4.56	less	−26.1
2014	−1.85	−0.52	−0.02	0.12	−0.16	−3.19	more	6.3
2015	0.62	−0.19	0.29	1.39	−0.22	2.72	more	11.7
2016	−0.49	−0.19	0.36	2.73	−0.36	0.72	less	9.5
2017	−2.48	0.48	0.03	2.31	−0.41	−1.66	more	−14.4
2018	3.79	1.14	0.03	2.14	−0.17	−4.5	less	−24.8

According to the prediction effect test from 2011 to 2018, the prediction accuracy of multi-factor synergy reached 87.5%, which was 25% higher than that of a single factor. In view of the fact that the analysis of the contemporaneous factors is more applied to diagnostic analysis, considering the actual situation of prediction, the SST index modeling of pre-winter (Figure 3) is selected to forecast the business according to the previous method.

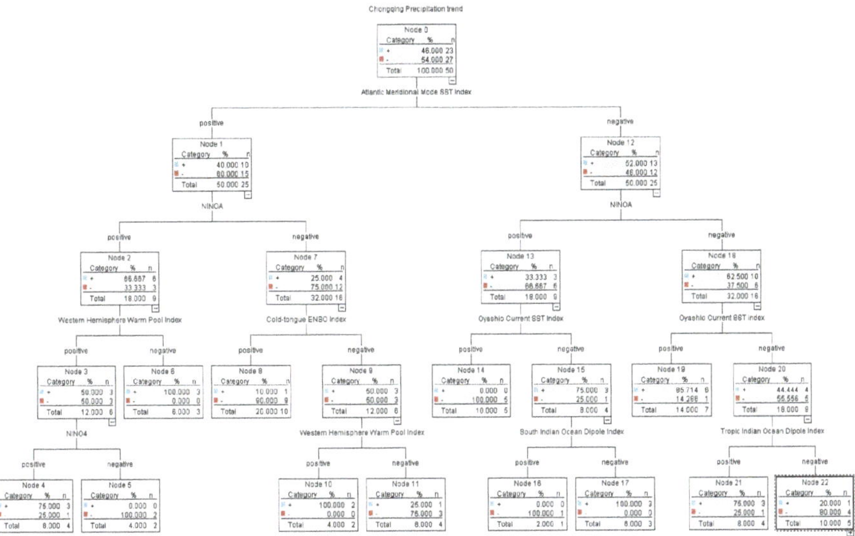

Figure 3. Relationship between precipitation trend in summer and SST index in pre-winter based on the CART algorithm in Chongqing.

The combination of summer precipitation trend and winter SST index model based on the CART algorithm in Chongqing is shown in Table 3. In the model, there are 6 cases of lower precipitation and 6 cases of higher precipitation.

The model was tested based on the observation of summer precipitation in Chongqing from 2011 to 2018, the results are shown in Table 4. In the model, if the Atlantic meridional model SST with the highest correlation is considered, the precipitation in Chongqing is low if it is high, while the precipitation in Chongqing is high if it is low. The trend forecast is correct in all years except 2014. If different combinations are considered, from 2011 to 2014, the Atlantic SST to mold is on the high side, and the NINOA is low. The cold tongue ENSO index was small and the predicted precipitation was small in 2013, which is consistent with the situation (1). The difference in the remaining three years is the difference in the western hemisphere warm pool index. In 2011 and 2012, it is consistent with the situation (2), with less predicted precipitation. In 2014 it is consistent with the situation (7), with more predicted precipitation. Signals of SST in 2015 and 2016 are consistent with the situation (10), and too much precipitation is predicted. In 2017 and 2018, it coincided with the situation (3), with less precipitation forecast. From the test, it can be seen that the forecast of precipitation trend in the 8 years from 2011 to 2018 is correct when considering the coordination of multiple factors, which is 12.5% higher than that when considering only a single factor.

The above considers the multi-factor synergy of the decision tree method. Although quantitative prediction of Chongqing's summer precipitation cannot be achieved, the experiments show that, no matter whether the predictive diagnosis analysis is made by using the previous or the same period factor, it is more obvious than the single index. This also shows that the "climate system", as a complex system, is the result of the interaction of multiple factors and multiple systems. In the process of diagnosis or prediction, we not only need to analyze the characteristics and cycles of each part of the system separately, but we must also study the integration behavior of the entire system and the interaction of the sub-system. This process requires statistical analysis of a large number of data such as ocean and atmosphere, as well as various model prediction data, in order to obtain the key factors affecting the local climate, the key regions of different circulation fields, and the key periods when indexes and circulation affect the local climate. With many "blind spots" in the physical processes and research of climate system change, current prediction methods cannot make full use of these huge data resources. It may be an important factor for large climate systems, but not necessarily a critical factor for local climatic characteristics. This will inevitably lead to "lighter and slightly heavier" situations in forecasting analysis, leading to uncertainty in the forecast Increased predictive accuracy. Therefore, with the help of decision tree and other machine learning technologies, comprehensive and valuable information can be fully mined from the vast variety of data, so as to discover the main system and collaborative influence mechanism that affect the local climate, which plays a significant role in improving the accuracy of local climate prediction.

3.2. Prediction Experiment of Random Forest Model in Summer

In the actual forecasting business, it is not only necessary to forecast the overall trend of the region, but also to analyze the spatial distribution pattern and forecast the rainfall centers and the occurrence locations. Therefore, based on the average model of the whole city in the previous section, this section uses random forests for prediction of 34 National Meteorological Observatories in Chongqing. In the selection of circulation index, since the actual summer forecast is released in March, the circulation factor that can be obtained at this time can only reach February. Thus, when random forests are used for prediction, this article only uses the early winter SST index modeling, regardless of the constraints such as the distribution conditions, interactions, nonlinear effects, and even missing values of variables. Figure 4 is the forecast distribution map of random forest precipitation and distribution of actual precipitation anomaly rate over the years 2011–2018.

It can be seen from Figure 4 that there was no consistent or excessive summer precipitation in Chongqing during 2011–2018, which is a case of different spatial distributions, which also makes prediction difficult. Comparing the forecast with the actual situation, the overall trend forecast for 8 years is more accurate. Only the spatial distribution of 2011 and 2015 is slightly different, and the remaining years are relatively accurate in regional forecast. Because the forecast uses a dichotomous trend forecast and cannot be refined for anomalous forecasting, the prediction results are tested at 20% and −20% using Ps, Cc and PC test methods, respectively. The test results are shown in Table 5.

As can be seen from Table 5, the random forest prediction score is higher and more stable. The average Ps, Cc and PC scores for 2014-2018 were respectively 84.6, 0.27 and 67.1. Compared with 72.4, −0.12, and 52.9, which are released by the forecast, they are significantly improved. From the historical comparison, Ps and PC scores are consistent. 2016 and 2017 are roughly equivalent to the released forecasts, and the rest of the years are about 20 points higher than the released forecasts. The Cc score for the correlation between the predicted field and the live field is significantly better than the forecast, and, except for 2015, they all exceed 95% significance test. In contrast, the published forecast shows that Cc scores are mostly negative, which indicates that the predictive typing needs to be improved.

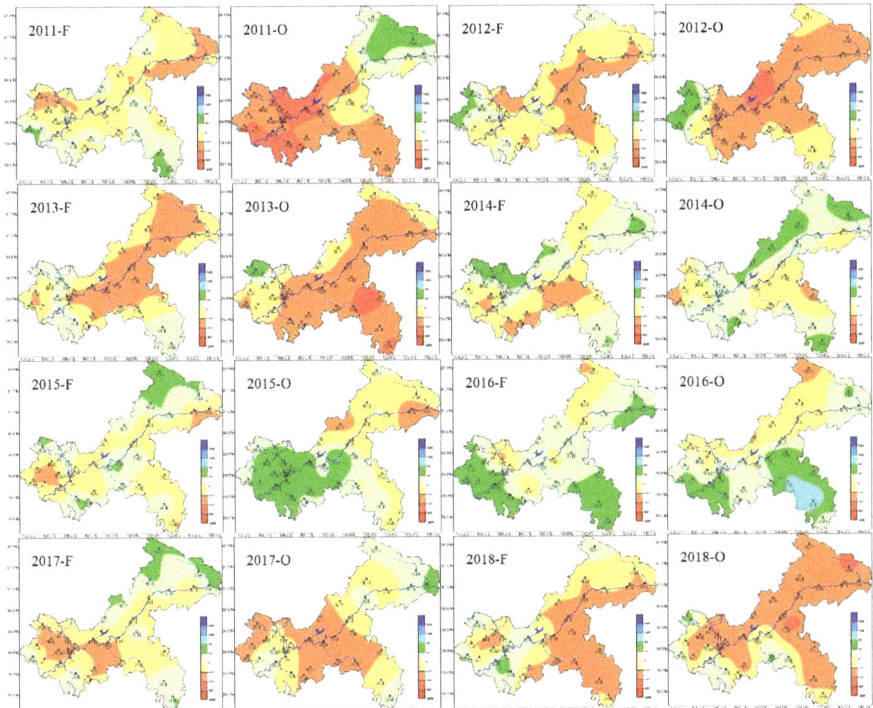

Figure 4. Forecast and observation distribution of summer precipitation in Chongqing based on random forest. (F) and (O) mean forecasting and observation, respectively.

Table 3. Combination of summer precipitation trend and prewinter Sea Surface Temperature (SST) model based on the CART algorithm in Chongqing.

Situation	Condition 1	Condition 2	Condition 3	Condition 4
1	High Atlantic meridional SST (+) (60%)	NINOA low "−" (75%)	Cold tongue ENSO index is larger "+" (90%)	
2	High Atlantic meridional SST (+) (60%)	NINOA low "−" (75%)	Cold tongue type ENSO index is small "−" (50%)	Warm pool index of the western hemisphere is relatively small "−" (75%)
3	High Atlantic meridional SST (+) (60%)	NINOA high "+" (33%)	Warm pool index of the western hemisphere is larger "−" (50%)	NINO4 low "−" (100%)
4	Low Atlantic meridional SST "−" (48%)	NINOA high "+" (67%)	High SST in tidophilic area "+" (100%)	
5	Low Atlantic meridional SST "−" (48%)	NINOA high "+" (67%)	Low SST in tidophilic area "−" (25%)	ositive dipole anomaly "+" (100%) in the subtropical southern Indian Ocean
6	Low Atlantic meridional SST "−" (48%)	NINOA low "−" (38%)	Low SST in tidophilic area "−" (56%)	Cold tongue type ENSO index is small "−" (80%)
7	High Atlantic meridional SST (+) (40%)	NINOA high "+" (67%)	The warm pool index of the western hemisphere is relatively small "−" (100%)	
8	High Atlantic meridional SST (+) (40%)	NINOA high "+" (67%)	Warm pool index of the western hemisphere is larger "+" (50%)	NINO4 high SST "+" (75%)
9	High Atlantic meridional SST (+) (40%)	NINOA low "−" (67%)	Cold tongue type ENSO index is small "−" (50%)	Warm pool index of western hemisphere is larger "+" (100%)
10	Low Atlantic meridional SST "−" (52%)	NINOA high "+" (33%)	Low SST in tidophilic area "−" (75%)	Sub-tropical southern Indian Ocean dipole is small "−" (100%)
11	Low Atlantic meridional SST "−" (52%)	NINOA low "−" (63%)	High SST in tidophilic area "+" (85%)	
12	Low Atlantic meridional SST "−" (52%)	NINOA low "−" (63%)	Low SST in tidophilic area "−" (44%)	Cold tongue ENSO index is larger "+" (75%)

Table 4. Test table for predicting summer precipitation effect from 2011 to 2018 based on the winter SST index decision tree model in Chongqing.

Years	Atlantic Meridional SST	NINOA	Western Hemisphere Warm Pool Index	Cold Tongue Type ENSO Index	SST in Tidophilic Zone	NINO4	Warm Pool Type ENSO Index	Subtropical Southern Indian Ocean Dipole	Tropical Indian Ocean Dipole	Prediction	Observation (%)
2011	6.04	−0.23	−0.1	−0.46	1.92	−0.94	−0.83	0.6	0.12	less	−30.5
2012	2.98	−0.09	−0.05	−0.5	0.96	−0.75	−0.34	−0.41	0.17	less	−22.1
2013	2.8	−0.57	0.14	0.16	1.45	−0.06	−0.77	−0.61	0.13	less	−26.1
2014	0.57	−0.32	0.29	−0.06	1.24	−0.25	−0.38	−0.53	−0.05	more	6.3
2015	−0.41	−0.33	0.16	0.49	−0.28	0.68	0.2	−0.26	−0.14	more	11.7
2016	−0.6	0.71	2.38	0.24	−0.52	1.2	1.91	−0.43	−0.28	more	9.5
2017	1.79	0.46	0.35	−0.29	−0.35	−0.37	−0.08	0.53	0.04	less	−14.4
2018	2.68	0.09	0.3	0.46	0.85	−0.94	−0.82	0.06	0.47	less	−24.8

Table 5. Comparison Table of random forest summer precipitation and release forecast from 2011 to 2018.

	Ps		Cc		PC	
Years	Random Forests	Report	Random Forests	Report	Random Forests	Report
2011	82.2		0.12		61.8	
2012	95.2		0.55		85.3	
2013	90.3		0.28		73.5	
2014	78.8	58.8	0.26	0.15	58.8	38.2
2015	79.4	58.2	0.03	−0.58	58.8	29.4
2016	86.1	83.9	0.31	−0.06	70.6	73.5
2017	85.0	85.7	0.30	0.25	64.7	67.6
2018	93.5	75.4	0.44	−0.37	82.4	55.9
2014–2018.	84.6	72.4	0.27	−0.12	67.1	52.9

4. Conclusions and Discussion

By establishing a decision tree model based on multi-factor collaboration for summer precipitation in Chongqing and conducting random forest integration and testing, the following conclusions are reached:

(1) In the concurrent circulation index that affects summer precipitation in Chongqing, the western Pacific subtropical ridge is a very important influencing factor. However, if we only consider the West Pacific sub-ridge ridgeline, there are a total 5 years in the 2011–2018 trend forecast which are accurate. Considering the synergistic effect of the Indian sub-high area and the typhoon landing, the 8-year trend can be accurately predicted, and the trend consistency rate increased by 37.5%. In the case that multiple factors were taken into account for the SST factor in the first winter, the precipitation trend prediction in 8 years was correct, which was 12.5% higher than that in the case that only a single factor of Atlantic meridional model SST was considered. This shows that in the prediction business, as the climate system is the result of the interaction of multiple factors and multiple systems, we not only need to analyze the characteristics and cycles of each part of the system separately, but also need to study the integration behavior of the whole system and the interaction of each subsystem. Using the decision tree to construct a multi-system collaborative impact model is an effective technical method. It is able not only to effectively improve the prediction accuracy, but also the prediction model established by the decision tree is different from the fully black-box effect of the neural network. The affected processes are relatively clear, so there is a higher application prospect for researches such as mechanism analysis.

(2) Using random forest to predict the summer precipitation Ps, Cc and PC scores of Chongqing from 2014 to 2018 are steadily higher than the released forecasts. In addition to the instability of publishing forecasts, the quality of random forest forecasts is relatively stable. The results show that it is feasible to use the random forest algorithm to predict summer rainfall precipitation in Chongqing in actual business. In addition, the random forest algorithm does not have high requirements on data, and it does not need to consider constraints such as the distribution conditions, interaction, nonlinear effects, even missing values of variables. In most cases, the default parameters of the model can give optimal simulation results without tedious parameter adjustment. Therefore, the application of the random forest algorithm in the climate prediction business has good prospects.

In this paper, when using decision trees and random forests to predict and model summer precipitation in Chongqing, although the model has a good prediction effect, it is also a qualitative forecast. Quantitative prediction modeling research has not been carried out, and there are obvious limitations in precipitation prediction and central locations. The author will increase the research and development of multi-factor collaboration, multi-system integration and multi-mode collection technology in subsequent research and business. Further analysis is made on various factors affecting summer rainfall in Chongqing, so as to provide more evidence and clues for improving the precipitation forecasting level in this region.

Author Contributions: Conceptualization, C.Z. and X.D.; methodology, C.Z., X.D. and B.X.; software, B.X.; validation, X.D., B.X. and C.Z.; formal analysis, X.D. and C.Z.; investigation, J.W.; resources, X.D.; data curation, J.W.; writing—original draft preparation, X.D., C.Z. and J.W.; writing—review and editing, X.D., C.Z. and J.W.; visualization, B.X.; supervision, J.W.; project administration, X.D.; funding acquisition, C.Z. and X.D. All authors have read and agreed to the published version of the manuscript.

Funding: This research was funded by national natural science foundation of China (41875111), Chongqing natural science foundation project (cstc2019jcyj-msxmX0227), Chongqing technology innovation and application demonstration general project (cstc2018jscx-msybX0165), Intelligent meteorological technology innovation team project of Chongqing meteorological bureau (ZHCXTD-201804), Data availability. The data in this study are not available for use by others.

Conflicts of Interest: The authors declare no conflict of interest.

References

1. Yong-Hua, L.I.; De, L.I.U.; Ye-Yu, Z.H.U.; Yang-Hua, G.A.O.; Wen-shu, M.A.O. Singular Spectrum Analysis of Surface Air Temperature and Precipitation Series in Chongqing. *Plateau Meteorol.* **2005**, *24*, 798–804.
2. Yong Hua, L.; Wen Shu, M.; Yang Hua, G.; Feng Qing, H.; Jia Qi, L. Regional Flood and Drought Indices in Chongqing and their Variation Features Analysis. *J. Meteorol. Sci.* **2006**, *26*, 638–644.
3. Li, Y.H.; Gao, Y.H.; Han, F.Q.; Xiang, M.; Tang, Y.H.; He, Y.K. Features of Annual Temperature and Precipitation Variety with the Effects on NPP in Chongqing. *J. Appl. Meteorol. Sci.* **2007**, *18*, 73–79.
4. Yi, Z.; Yanghua, G.; Xionghong, D. Primary Climatic Characteristics of Summer Precipitation in the Three-Gorges Reservoir Region. *J. Southwest Univ. (Nat. Sci. Ed.)* **2005**, *27*, 269–272.
5. Zhenfeng, M. Forecast of Summer Precipitation over Southwest Region of China. *Meteorol. Mon.* **2002**, *28*, 29–33.
6. Zhang, Q.; Jiang, T.; Wu, Y.J. Impact of ENSO Events on Flood/Drought Disasters of Upper Yangtze River during 1470–2003. *J. Glaciol. Geocryol.* **2004**, *26*, 691–696.
7. De, L.; Yong-hua, L.I.; Yang-hua, G.A.O.; Jing, L.I.; Yun-hui, T.A.N.G.; Zhao, Y.E. Analysis on Eurasian Circulation of Drought and Flood in Summer of Chongqing. *Plateau Meteorol.* **2005**, *24*, 275–279.
8. Xiu, Y.Y.; Han, L.; Feng, H.L. The identification of strong convective weather based on machine learning methods. *Electron. Des. Eng.* **2016**, *24*, 4–7.
9. Quande, S.; Ruili, J.; Jiangjiang, X.; Zhongwei, Y.; Haochen, L.; Jianhua, S.; Lizhi, W.; Zhaoming, L. Adjusting Wind Speed Prediction of Numerical Weather Forecast Model Based on Learning Methods. *Meteorol. Mon.* **2019**, *45*, 426–436.
10. Li, W.; Zhao, F.; Li, M.; Chen, L.; Peng, X. Forecasting and Classification of Severe Convective Weather Based on Numerical Forecast and Random Forest Algorithm. *Meteorol. Mon.* **2018**, *44*, 1555–1564.
11. Jones, N. How machine learning could help to improve climate forecasts. *Nature* **2017**, *548*, 379–380. [CrossRef] [PubMed]
12. Huang, R.F.; Zhou, G.C. *Meteorology and Big Data*; Science Press: Beijing, China, 2017.
13. Zhao, Z.Y. *Python Machine Learning Algorithm*; Electronic Industry Press: Beijing, China, 2017.
14. Shi, D.; Geng, H.; Ji, C.; Huang, C. Construction and application of road icing prediction model based on C4.5 decision tree algorithm. *Meteorol. Sci.* **2015**, *35*, 204–209.
15. Shi, Y.; Shi, D.; Hao, L.; Zhang, Y.; Wang, P. Research on classification and prediction model of regional summer precipitation days based on CART algorithm of data mining. *J. Nanjing Univ. Inf. Technol. (Nat. Sci. Ed.)* **2018**, *10*, 118–123.
16. Qin, P.C.; Liu, Z.X.; Wan, S.Q.; SU, R.R.; Huang, J.F. Yield limiting factor analysis of rapeseed in Hubei province based on decision tree and random forest model. *Chin. J. Agrometeorol.* **2016**, *37*, 691–699.
17. Zhang, R.; Zhang, R.; Zuo, Z. Impact of Eurasian spring snow decrement on East Asian summer precipitation. *J. Clim.* **2017**, *30*, 3421–3437. [CrossRef]
18. Wu, B.; Su, J.; D'Arrigo, R. Patterns of Asian winter climate variability and links to arctic sea ice. *J. Clim.* **2015**, *28*, 6841–6858. [CrossRef]
19. Weng, H.; Wu, G.; Liu, Y.; Behera, S.K.; Yamagata, T. Anomalous summer climate in China influenced by the tropical Indo-Pacific Oceans. *Clim. Dyn.* **2011**, *36*, 769–782. [CrossRef]
20. Yuan, Y.; Yang, S.; Zhang, Z. Different evolutions of the Philippine Sea anticyclone between eastern and central Pacific El Niño: Possible effect of Indian Ocean SST. *J. Climate* **2012**, *25*, 7867–7883. [CrossRef]
21. Wei, W.; Fengchang, X.; Dawei, S.; Xiaojie, S. Research and application of CART algorithm based summer drought prediction model. *J. Meteorol. Sci.* **2016**, *36*, 661–666.
22. Breiman, L. Random forests. *Mach. Learn.* **2001**, *45*, 5–32. [CrossRef]
23. Wu, J.; Chen, Y.F.; Yu, S.N. Research on drought prediction based on random forest model. *China Rural Water Resour. Hydropower* **2016**, *11*, 17–22.
24. Binren, X.U.; Yuanyuan, W.E.I. Spatial statistical reduction of precipitation data of TRMM on Qinghai-Tibet plateau based on random forest algorithm. *Remote Sens. Land Resour.* **2018**, *30*, 181–188.
25. Qingquan, L.; Yiming, D.; Yihui, L. 10-Year Hindcasts and Assessment Analysis of Summer Rainfall over China from Regional Climate Model. *J. Appl. Meteorol. Sci.* **2005**, *S1*, 41–47.

26. Guang-tao, D.; Bo-min, C.; Bao-de, C. Application of Regional Climate Model (RegCM3) on 10-Year Hindcast Experiment and a Real-Time Operation in Summer of 2010 in the Eastern China. *Plateau Meteorology.* **2012**, *31*, 1601–1610.

27. Bai, H.; Gao, H.; Liu, C.Z.; Mao, W.Y.; Du, L.M. Assessment of Multi-model Downscaling Ensemble Prediction System for Monthly Temperature and Precipitation Prediction in GuiZhou. *Desert Oasis Meteorol.* **2016**, *10*, 58–63.

28. Yang, X.B.; Zhang, J. Decision Tree and Its Techniques. *Comput. Technol. Dev.* **2007**, *17*, 43–45.

29. Quinlan, J.R. *C4.5: Programs for Machine Learning*; Morgan Kaufman: San Mateo, CA, USA, 1993.

30. Iverson, L.R.; Prasad, A.M.; Matthews, S.N.; Peters, M. Estimating potential habitat for 134 eastern US tree species under six climate scenarios. *For. Ecol. Manag.* **2008**, *254*, 390–406. [CrossRef]

31. Wang, W.J.; Yao, Z.Y.; Jia, S.; Zhao, W.H.; Tan, C.; Zhang, P.; Gao, L.S.; Zhu, X.Y. Application Research on Random Forest Algorithm in the Statistical Test of Rainfall Enhancement Effect. *Meteorol. Environ. Sci.* **2018**, *41*, 111–117.

32. Men, X.L.; Jiao, R.L.; Wang, D.; Zhao, C.G.; Liu, Y.K.; Xia, J.J.; Li, H.C.; Yan, Z.W.; Sun, J.H.; Wang, L.Z. A temperature correction method for multi-model ensemble forecast in North China based on machine learning. *Clim. Environ. Res.* **2019**, *24*, 116–124. (In Chinese)

33. Gao, H.; Ding, T.; Li, W. The three-dimension intensity index for western Pacific subtropical high and its link to the anomaly of rain belt in eastern China (in Chinese). *Chin. Sci. Bull.* **2017**, *62*, 3643–3654. [CrossRef]

34. Shao, X.; Zhou, B. Monitoring and Diagnosis of the 2015/2016 Super El Nino Event. *Meteorol. Mon.* **2016**, *42*, 540–547.

Article

The Contribution Rate of Driving Factors and Their Interactions to Temperature in the Yangtze River Delta Region

Cheng Zhou [1], Nina Zhu [2,3,4,*], Jianhua Xu [2,3,4,*] and Dongyang Yang [5]

1   Faculty of Culture and Tourism, Shanxi University of Finance and Economics, Taiyuan 030006, China
2   Key Laboratory of Geographic Information Science (Ministry of Education), East China Normal University, Shanghai 200241, China
3   Research Center for East–West Cooperation in China, East China Normal University, Shanghai 200241, China
4   School of Geographic Sciences, East China Normal University, Shanghai 200241, China
5   Key Research Institute of Yellow River Civilization and Sustainable Development, Henan University, Kaifeng 475004, China
*   Correspondence: ninaecnu@126.com (N.Z.); jhxu@geo.ecnu.edu.cn (J.X.);
    Tel.: +86-1592-186-5239 (N.Z.); +86-139-1834-3871 (J.X.)

Received: 6 November 2019; Accepted: 24 December 2019; Published: 27 December 2019

Abstract: Complex temperature processes are the coupling results of natural and human processes, but few studies focused on the interactive effects between natural and human systems. Based on the dataset for temperature during the period of 1980–2012, we analyzed the complexity of temperature by using the Correlation Dimension (CD) method. Then, we used the Geogdetector method to examine the effects of factors and their interactions on the temperature process in the Yangtze River Delta (YRD). The main conclusions are as follows: (1) the temperature rose 1.53 °C; and, among the dense areas of population and urban, the temperature rose the fastest. (2) The temperature process was more complicated in the sparse areas of population and urban than in the dense areas of population and urban. (3) The complexity of temperature dynamics increased along with the increase of temporal scale. To describe the temperature dynamic, at least two independent variables were needed at a daily scale, but at least three independent variables were needed at seasonal and annual scales. (4) Each driving factor did not work alone, but interacted with each other and had an enhanced effect on temperature. In addition, the interaction between economic activity and urban density had the largest influence on temperature.

Keywords: correlation dimension method; Geogdetector method; interaction effect; multi-scale

---

1. Introduction

The climate system is a complex system, which influence on ecosystems and human society has attracted more and more attention from scholars, and temperature is one of the most important factors in the climate system.

A number of studies [1–7] have indicated that the spatiotemporal variation of temperature and its driving factors had regional differences. Sharma et al. [8] analyzed the temperature changes in eastern India, and the results showed that the average temperature in central, southern and western was decreasing, while the average temperature in the northeast, west and southeast was on the rise. Salman et al. [9] conducted a hybrid model to select the climate models for simulating spatiotemporal changes in temperature of Iraq. They found that temperature would increase during the period of 2070–2099 and temperatures in the north and northeast had increased significantly. Kenawy et al. [10] pointed out that temperatures in northeastern Spain showed an upward trend during the 1960–2006 period, and the Eastern Atlantic (EA), the Scandinavian (SCA), and the Western Mediterranean Oscillation (WeMO) patterns had a significant impact on temperature changes. Iqbal et al. [11] found temperature had different correlations with North Atlantic Oscillation (NAO), Arctic Oscillation (AO), El Niño-Southern Oscillation (ENSO), and North Sea Caspian Pattern (NCP) in different months in Pakistan. It can be seen that the temperature change is complicated. Therefore, further understanding the mechanisms for spatio-temporal variation of temperature and its driving factors are highly desired.

In order to reveal the complexity of temperature, many methods had been proposed, such as wavelet analysis [12,13], ensemble experience mode decomposition [14], spectrum analysis [15], Mann-Kendall trend test [16], and correlation dimension [17]. All of these methods had explored the complexity of temperature from different perspectives and got some achievements. On the other hand, there are many studies about the driving factors of temperature change. The main driving factors are atmospheric circulation [2], land use changes [3], greenhouse gas emissions [18], urbanization development [19], and so on. However, under the global warming, coupled with rapid economic development, population growth, and urbanization, the temperature and its driving factors became more and more complicated. In addition, the contribution rate of natural and socioeconomic factors and their interactions on temperature variation were rarely studied and remained one of main gaps in our current knowledge.

Due to the regional differences, it is necessary to conduct an in-depth analysis of temperature variations in some key areas, especially those that play an important role in national development. The Yangtze River Delta (YRD), one of China's most developed, dynamic, densely populated and concentrated industrial areas, is growing into an influential world-class metropolitan area. However, the developed industries and frequent human activities have led to an increasingly serious urban heat island phenomenon in this region, forming a strong regional heat island, leading to the temperature presenting a significant warming trend over the past 50 years and extremely high temperatures occuring frequently [20]. Property, economic losses, and social impacts caused by extremely temperature events in this region are often enormous. In addition, extreme changes in temperature can also have an impact on the environment and endanger human health [21]. Therefore, it is of great significance to study the temperature changes in this region, to find the reasons that affect its changes, and to try to reduce losses.

Therefore, we attempt to explore the spatiotemporal dynamics of temperature in the YRD, and assess the influences of factors and their interactions on temperature. Based on observed temperature data at 68 meteorological stations during the period of 1980–2012, we first investigated the spatiotemporal complexity of temperature by using the Correlation Dimension (CD) method; and then we analyzed the individual contribution rates and interactional contribution rates of driving factors to temperature slope (TS) by using the Geogdetector method. Our main purpose is to explore which factors or interactions between factors contribute the most to temperature.

2. Materials and Methods

2.1. Study Area and Data

The study area includes four regions: Jiangsu Province, Anhui Province, Zhejiang Province, and Shanghai (Figure 1). The study area lies between 114°54′–122°42′ E and 27°12′–35°20′ N, and has an area of approximately 344.03 10^3 km^2, accounting for 3.58% of China's total land area. The area is under a monsoon climate regime, with hot and humid summer and cold and dry winter. The annual precipitation is about 1000 mm, of which the precipitation in summer accounts for two-thirds of the total precipitation [22]. The average temperature is close to 30 °C in July and August, and the maximum temperature recently exceeded 40 °C in Shanghai [23]. The high terrain is in the north and south and low terrain in the middle, which is dominated by plains and hills. In addition, the YRD is one of the most developed regions in China, with dense population, convenient transportation, and developed tertiary industry.

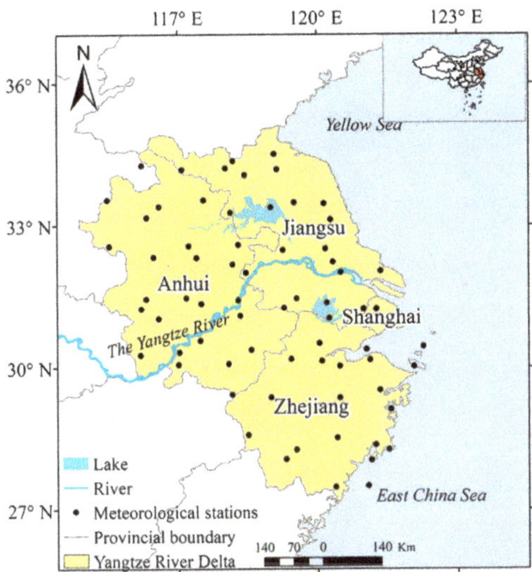

Figure 1. The study area and spatial distribution of 68 meteorological stations.

On a global scale, the temperature is mainly affected by factors such as atmospheric circulation, volcanic eruptions, sunspots, and so on. However, on the regional scale, the temperature is mainly affected by surface properties and human activities. According to previous studies [24–26], the altitude (AT), normalized difference vegetation index (NDVI), urban density (UD), gross domestic product (GDP), and night light (NL) datasets were selected. The first two can be seen as natural factors and the last three can be seen as socioeconomic factors. The daily temperature of 68 meteorological stations from 1980 to 2012 is from the China Meteorological Data Service Center (http://data.cma.com). We analyzed data from the period from 1980 to 2012 because we couldn't get station data of temperature for 2013–2018. AT and UD data are provided by the Data Center for Resources and Environmental Sciences, Chinese Academy of Sciences (RESDC) (http://www.resdc.cn), and the UD data is from 1990 to 2010. NDVI is from the Geospatial data cloud (http://www.gscloud.cn/), and its period is from 1989 to 2012. NL is from the National Centers for Environmental Information (https://www.ngdc.noaa.gov/), and its period is from 1992 to 2012. GDP is from the "Shanghai Statistical Yearbook", "Anhui Statistical Yearbook", "Jiangsu Statistical Yearbook", "Zhejiang Statistical Yearbook", and "China Regional Economic Statistics Yearbook" and other statistics, and its period is from 1980–2012. To ensure a consistent data format, a 0.5 km by 0.5 km grid for the whole area in ArcGIS 10.5 software (Manufacturer, City, US State abbrev. if applicable, Country) was built, assigned values to each grid, and deleted the outliers by using a box-plot analysis method. According to different standards, all factors were divided into different strata using ArcGIS 10.5 software. The division of the results is shown in Figure 2 below.

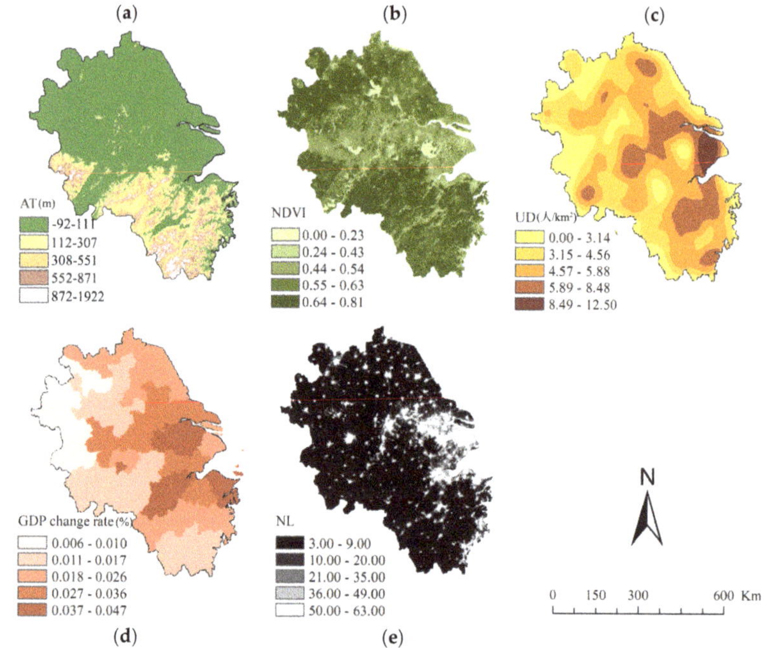

Figure 2. The distribution of driving factors: (**a**) altitude (AT); (**b**) normalized difference vegetation index (NDVI); (**c**) urban density (UD); (**d**) gross domestic product (GDP) change rate; (**e**) night light (NL).

2.2. Methods

To investigate the spatiotemporal complexity of temperature and its driving factors, the correlation dimension method and the Geogdetector method were used. It can be seen from Figure 3, we first showed the spatiotemporal pattern of temperature; then, we analyzed the complexity of temperature on the daily, seasonal, and annual scales by using the Correlation Dimension (CD) method;finally, the individual contribution rates and interactional contribution rates of driving factors to the temperature slope (TS) by using Geogdetector method were detected.

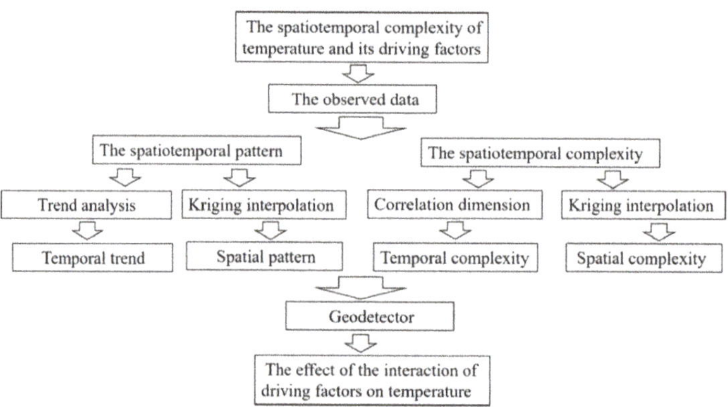

Figure 3. The framework of this study.

2.2.1. Trend Analysis Method

Trend analysis is the most studied and most popular quantitative forecasting method by far. It is based on a known historical data to fit a curve, so that this curve can reflect the growth trend of things themselves, and then to predict the future according to this growth trend curve. Commonly used trend models include linear trend models, polynomial trend models, linear trend models, log trend models, power function trend models, exponential trend models, and so on [27]. In this study, we use the linear trend method to analyze the change trend of the time series:

$$y(t) = at + b,\tag{1}$$

where y represents the time series, t represents the time, a represents the linear slope, and b represents the intercept.

If $a > 0$, it indicates that the time series is increasing, if $a = 0$, it means that the time series is not changing, and, if $a < 0$, it indicates that the time series is decreasing. The size of a indicates the degree of change in time series.

2.2.2. Kriging Interpolation Method

Temperature is a regionalized variable, which is changing with the variation of space position. In order to analyze the distribution of the plum rainfall in different years, Kriging interpolation is employed. Kriging interpolation (or space local estimation) is named by D. G. Krige, who is a mining engineer in South Africa, and it is an optimal interpolation method [28]. The original data of the regional variables and the structural characteristics of the variance function is used to estimate the value of non-sampling points unbiasedly and optimally [28]. In general, Kriging interpolation contains several types, namely Ordinary Kriging, Universal Kriging, Co-Kriging, and so on. Ordinary Kriging is shown below:

Assume that $Z(x)$ is a regionalized variable that satisfies two-stage stationary hypotheses and intrinsic hypothesis. m is mathematical expectation, with covariance function and variance function all existing at the same time. The relation between them is indicated below:

$$E[Z(x)] = m,\tag{2}$$

$$C(h) = E[Z(x)Z(x+h)] - m^2,\tag{3}$$

$$\gamma(h) = \frac{1}{2}E[(Z(x) - Z(x+h)]^2.\tag{4}$$

Assuming that there are *no* measured points in the neighborhood of x, namely $x_1, x_2, \ldots, x_n$, for which the sample value is $Z(x_i)(i = 1, 2, 3, \ldots, n)$, the formula can be defined as follows:

$$Z^*(x) = \sum_{i=1}^{n} \lambda_i Z(x_i),\tag{5}$$

where λ_i is a weight coefficient that presents the contribution degree of the observed values of $Z(x_i)$ to estimate the values of $Z^*(x)$. Two points need to be noticed about this formula: on the one hand, the estimated value of $Z^*(x)$ must be unbiased, namely the mathematical expectation of the deviation is zero; on the other hand, it must be optimal, namely the difference between the estimated value and the actual value is the smallest.

2.2.3. Correlation Dimension (CD)

Since the appearance of fractal theory, fractal dimension has been welcomed by scholars as one of the quantitative indicators to describe whether the dynamic system has chaotic characteristics. There are different types of fractal dimensions, such as topological dimension, Hausdorff dimension, information dimension, and correlation dimension. As for the correlation dimension, Grassberger and Procaccia [29] proposed an analysis method for experimental time series data in 1983, which is to obtain the fractal dimension through the relationship between the integral $C\ (r)$ and the distance r on the reconstructed phase space through the univariate time series. This method is called a G–P method. Because it is particularly suitable for experimental observation data and the algorithm is simple and easy to implement, it has been widely used. In this study, we use this method when calculating the correlation dimension.

The correlation dimension (CD) is usually applied to analyze time series and determine if it exhibits a chaotic dynamic characteristic [30,31]. Considering $\{x_1, x_2, x_3, \ldots , x_i, \ldots \}$, the equal interval time series of daily temperature, and the first m data are extracted, and they determine the first point in the m-dimensional space, which is denoted as X_1. Then, remove x_1, and take m data $x_2, x_3, \ldots ,$ x_{m+1}, and the second point is composed of this set of data in m-dimensional space, which is recorded as X_2. According to this, a series of phase points $X_1, X_2, \ldots , X_N$ are formed. Given the number r, and check how many point pairs (X_i, X_j) distance is less than r, and the ratio of the number that point pairs distance is less than r to the total number of point pairs N is denoted as $C(r)$ [17]. It can be expressed as the following formula:

$$C(r) = \frac{1}{N^2} \sum_{\substack{i,j \neq 1 \\ i \neq j}}^{N} \theta\left(r - \left|X_i - X_j\right|\right), \tag{6}$$

where $\theta(x)$ is the Heaviside function, which is defined as:

$$\theta(x) = \begin{cases} 1 & , & x > 0 \\ 0 & , & x < 0 \end{cases}. \tag{7}$$

If r is too large, the distance of all point pairs will not exceed it. In addition, this r cannot measure the correlation between phase points. In addition, appropriate reduce r, the following formula may exist:

$$C(r) \propto r^d. \tag{8}$$

If this relationship exists, d is a dimension called the correlation dimension, denoted as D_2:

$$D_2 = \lim_{r \to 0} \frac{\ln C(r)}{\ln r}. \tag{9}$$

The limit here mainly represents a direction in which r is reduced, and it is not mean that the r must be close to 0. In the scale transformation of the actual system, there are scale restrictions in the magnitude of both directions. Exceeding this limit is beyond the featureless scale area.

Figure 4 shows the results of $\ln(r)$ versus $\ln(r)$, and the correlation dimension (d) versus embedding dimension (m) used the measured data of temperature in this paper. It is apparent that the correlation dimension, D_2, is given by the slope coefficient of $\ln(r)$ versus $\ln r$. According to $(\ln r, \ln C(r))$, D_2 can be obtained by the least squares method using a log–log grid (as shown in Figure 4a).

To detect the chaotic behavior of the system, the correlation dimension has to be plotted as a function of the embedding dimension (as shown in Figure 4b).

The MATLAB 2014a software (Manufacturer, City, US State abbrev. if applicable, Country) was used to calculate Correlation Dimension. First, we calculated the time series of daily average temperature, monthly average temperature, and annual average temperature of each meteorological station during the period of 1980–2012, and then calculated the CD value of each station on different time scales through programming using MATLAB software (The MathWorks, Natick, MA, USA).

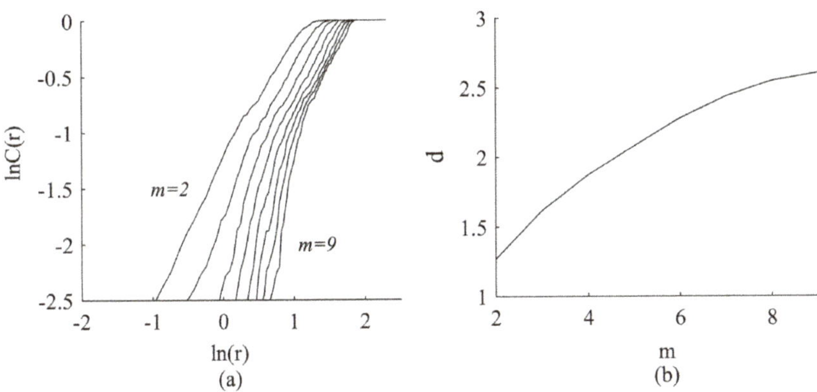

Figure 4. (**a**) a plot of ln(r) versus ln(r) and (**b**) the correlation dimension (*d*) versus embedding dimension (*m*).

2.2.4. Geogdetector

The influencing factors have spatial heterogeneity and work together to affect the temperature. Geogdetector is a set of statistical methods for detecting spatial variability and revealing forces driving the variability [32,33]. The advantages of this method are that it cannot only detect both quantitative and qualitative data, but also can detect the interaction of two factors [34]. Geogdetector contains four detectors: a risk detector, a factor detector, an ecological detector, and an interaction detector.

The risk detector can determine whether there is a significant difference in the means of attributes between two sub-regions, using the *t* statistic to test:

$$t_{\overline{Y}_{h=1}\,\overline{Y}_{h=2}} = \frac{\overline{Y}_{h=1} - \overline{Y}_{h=2}}{\left[\frac{Var\left(\overline{Y}_{h=1}\right)}{n_{h=1}} + \frac{Var\left(\overline{Y}_{h=2}\right)}{n_{h=2}}\right]^{\frac{1}{2}}},\tag{10}$$

where $\overline{Y}_h$ is the attribution average in the region h, n_h is the sample size of sub-region h, and *Var* is the variance. The statistic *t* approximates the Student's distribution, where the degree of freedom (*df*) is calculated as:

$$df = \frac{\frac{Var\left(\overline{Y}_{h=1}\right)}{n_{z=1}} + \frac{Var\left(\overline{Y}_{h=2}\right)}{n_{z=2}}}{\frac{1}{n_{h=1}-1}\left[\frac{Var\left(\overline{Y}_{h=1}\right)}{n_{h=1}}\right]^2 + \frac{1}{n_{h=2}-1}\left[\frac{Var\left(\overline{Y}_{h=2}\right)}{n_{h=2}}\right]^2}.\tag{11}$$

Null hypothesis: $\overline{Y}_{h=1} = \overline{Y}_{h=2}$. If H$_0$ is rejected at significance level α, there is a significant difference in the mean of the attributes between the two sub-regions.

The factor detector mainly detects the spatial variability of *Y* and the extent to which *X* is probed to explain the spatial differentiation of *Y*. The *q*-value was used to measure the factors:

$$q = 1 - \frac{\sum_{h=1}^{L} N_h \sigma_h^2}{N\sigma^2} = 1 - \frac{SSW}{SST},\tag{12}$$

$$SSW = \sum_{h=1}^{L} N_h \sigma_h^2, \quad SST = N\sigma^2,\tag{13}$$

where $h = 1, \dots, L$ is the stratum of *Y* or *X*, N_h, and *N* are the unit numbers of layer *h* and the unit numbers of the whole region, respectively, and σ_h^2 and σ^2 are the variances of *Y* of the layer *h* and of the whole region, respectively. *SSW* and *SST* are the sum of squares and the total sum of squares, respectively. The range of q is [0, 1]. The larger the value, the more obvious the spatial distribution of Y is. If the stratum is generated by the independent variable X, a larger q value shows stronger explanatory power of the independent variable X to Y, and a smaller q means weaker power. In extreme cases, a q value of 1 indicates that factor X has complete control over the spatial distribution of Y, and a q value of 0 indicates that factor X has no control over the spatial distribution of Y.

The ecological detector explores whether a geographical stratum, C, is more significant than another stratum, D, in controlling the spatial pattern, and the statistic F is used to measure it:

$$F = \frac{N_{X1}(N_{X2}-1)SSW_{X1}}{N_{X2}(N_{X1}-1)SSW_{X2}}, \tag{14}$$

$$SSW_X = \sum_{h=1}^{L1} N_h \sigma_h^2, \; SST_{X2} = \sum_{h=1}^{L2} N_h \sigma_h^2, \tag{15}$$

where N_{x1} and N_{x2} are the sample sizes of factors $X1$ and $X2$, respectively, and SSW_{x1} and SSW_{x2} are sums of the variances in the strata formed by $X1$ and $X2$, respectively. $L1$ and $L2$ represent the number of variables in $X1$ and $X2$, respectively. H_0 is $SSW_{x1} = SSW_{x2}$. If H_0 is rejected at the significance level of α, there is a significant difference in the spatial distribution of Y between $X1$ and $X2$.

The interaction detector is used to evaluate whether $X1$ and $X2$ together will increase or decrease the explanatory power of the dependent variable Y, or whether the effects of these factors on Y are independent of each other. $q(X1)$, $q(X2)$, and $q((X1 \cap X2)$ were calculated and compared the differences between $q(X1)$, $q(X2)$, and $q((X1 \cap X2)$.

The Geogdetector software was used to calculate Geogdetector. First, we need to calculate the annual average of temperature, AT, NDVI, UD, GDP change rate, and NL in their respective time periods, and convert the data format to .tif format; secondly, a 0.5×0.5 km grid is established by ArcGIS software, and each variable is extracted by the points in the grid; next, the extracted AT, NDVI, UD, GDP change rate, and NL were classified respectively. In this study, these variables are divided into five categories according to the natural segmentation method in ArcGIS software. Finally, the processed data is imported into the Geogdetector software for calculation.

3. Results

3.1. The Spatiotemporal Pattern of Temperature

In order to understand the temperature variations during the period of 1980–2012 in the YRD, we first analyzed the overall trend of temperature variation by using the linear trend method, and the linear slope is used to identify the trend of temperature changes. If the linear slope is greater than 0, it indicates that temperature is increasing, if the linear slope is equal to 0, it means that temperature is not changing, and, if the linear slope is less than 0, it indicates that temperature is decreasing. It showed a significant increasing trend during the period of 1980–2012 (Figure 5), and this trend may continue in the future. We can see that the temperature rose 1.53 °C with the average rising rate of 0.465 °C/10 years and passed the significance test during the period of 1980–2012. However, in the most recent 50 years, the global average rising rate only has reached about 0.13 °C/10 years [35]. We can conclude that the increase in temperature in the YRD was not only the result of global warming, but also other regional factors.

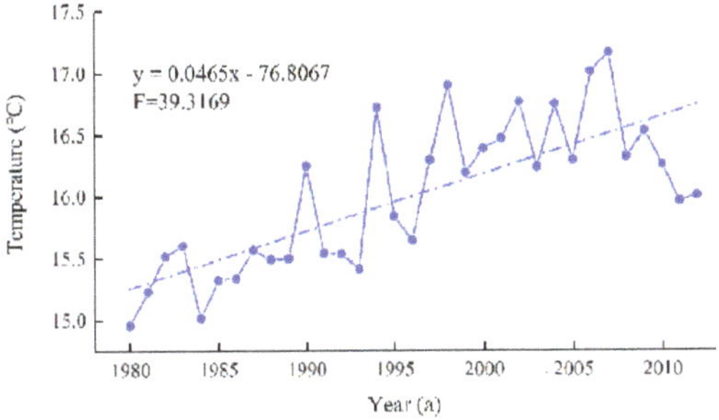

Figure 5. The trend of temperature during the period of 1980 to 2012.

Then, we showed the spatial distribution of TS during the period of 1980–2012 (Figure 6). The ordinary Kriging method was used during the interpolation, in which a spherical model is used when selecting the semi-variogram model, and its parameters are system default parameters. We can see that the temperature rose in all regions. In addition, among the dense areas of population and urban, the temperature rose quickly, while the temperature in the sparse areas of population and urban rose slowly.

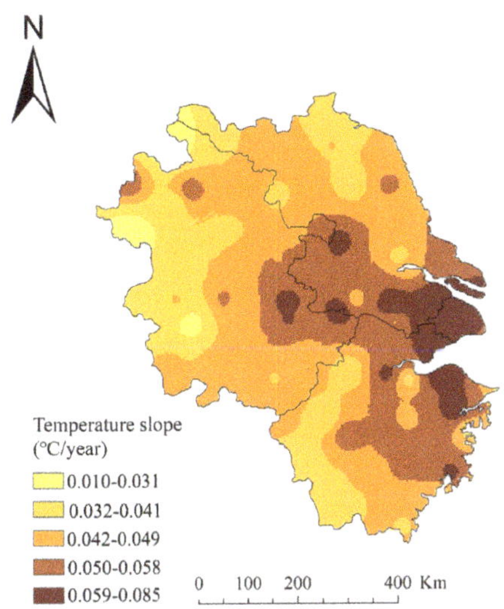

Figure 6. Temperature slope (change rate of temperature) during the period of 1980–2012.

3.2. The Spatiotemporal Complexity of Temperature

3.2.1. The Temporal Complexity of Temperature

Based on meteorological data, we analyzed the chaotic dynamics with fractal characteristic for the temperature dynamics by using the G–P method [36]. Firstly, we randomly selected six meteorological stations (i.e., Bozhou, Nanjing, Nantong, Wuhu, Hangzhou, Dongtou) with annual time series data for a pilot study. The relationship between different embedding dimension (m) and correlation exponent (d) was shown in Figure 7.

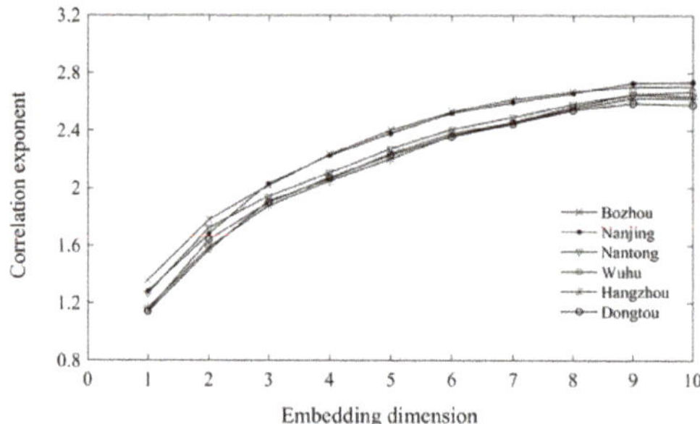

Figure 7. The plots of correlation exponent (d) versus embedding dimension (m) for the time series of annual data from the selected six meteorological stations

It can be seen from the trend of the six meteorological stations in Figure 7 that, as the embedding dimension increases, the correlation exponent increases continuously and eventually stabilizes, and the saturated correlation exponent, namely, the correlation dimension, was obtained when $m \geq 10$.

Then, we calculated the CD on the daily, seasonal and annual temporal scales of each station in the same way. Table 1 shows the CD values of several representative stations and average CD values of all stations at different temporal scales. It can be seen from Table 1 that each CD is not an integer, which indicates that the temperature process at different temporal scales is a chaotic dynamic system with a fractal characteristic, and it is sensitive to the changes of initial conditions.

Table 1. The Correlation Dimension (CD) values at daily, seasonal, and annual scales for 68 meteorological stations.

Station	Temporal Scale		
	Daily	Seasonal	Annual
Xuzhou	1.79	2.18	2.80
Fuyang	1.82	2.25	2.99
Nanjing	1.78	2.12	2.23
Nantong	1.62	2.02	2.11
Hefei	1.84	2.13	2.39
Baoshan	1.78	1.78	2.40
Huangshan	1.62	1.77	2.30
Hangzhou	1.69	1.86	2.05
Cixi	1.66	1.66	1.94
Jinhua	1.86	1.96	2.00
MCD	1.73	2.08	2.32

Note: MCD is the mean of correlation dimensions for all meteorological stations.

It can be seen from the mean of correlation dimensions (MCD) at different temporal scales in Table 1 that the ordering of the CD is: annual (2.32) > seasonal (2.08) > daily (1.73). We can conclude that the temperature process over a larger temporal scale is more complicated than the temperature process at a small temporal scale. Figure 8 showed the temperature anomalies of daily range and temperature anomalies of annual range. It could be seen from the maximum, minimum, and variance that the annual temperature fluctuated greatly, which proved that the temperature process on the annual scale was more complicated. Table 1 also shows that, even at the same temporal scale, the CD values of different stations are different. It is mainly related to the different locations of each station, which makes the driving factors of each station different. The values of MCD on the seasonal and annual scales are greater than 2, with 2.2 and 2.4, respectively, indicating that at least three independent variables are needed to describe the dynamics of temperature process on the seasonal and annual scale;

and the value of MCD for daily is 1.73, indicating that at least two independent variables are needed to describe the dynamics of temperature process on the daily scale.

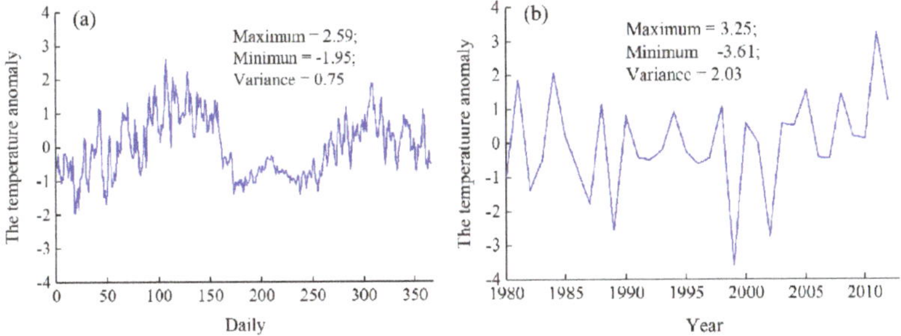

Figure 8. (a) anomalies of the daily range and (b) anomalies of the annual range.

3.2.2. The Spatial Distribution Complexity of Temperature

Table 1 gives the CD values of the temperature on different temporal scales, showing temperature dynamics on the daily, seasonal and annual scales. What is the spatial distribution of the CD values of different stations? We show the spatial distribution of CD values on the daily, seasonal, and annual scales (Figure 9). The ordinary Kriging method was used during interpolation, in which a spherical model is used when selecting the semi-variogram model, and its parameters are system default parameters.

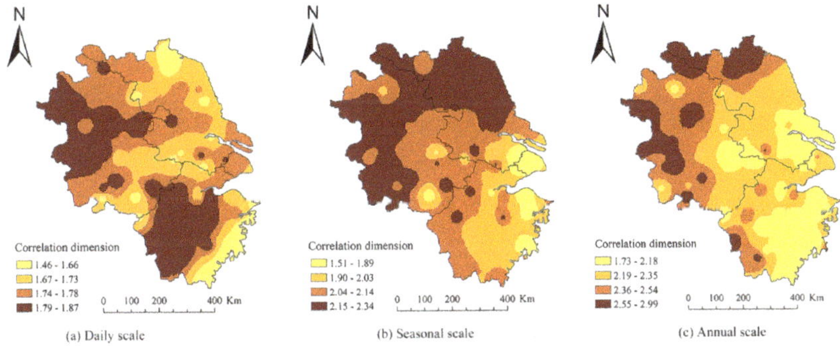

Figure 9. The spatial pattern of complexity of the temperature process at daily (**a**), seasonal (**b**), and annual (**c**) scales.

Figure 9a shows the spatial distribution of CD values on the daily scale, with values between 1.46 and 1.87. High value is mainly distributed in the northwest and southwest of the entire region, while low values are mainly distributed in the eastern coastal areas. Figure 9b presents the spatial distribution of CD values on the seasonal scale, which shows that all CD values are between 1.51 and 2.34. High value is mainly distributed in the northwest of the entire region, while low values are mainly distributed in the eastern coastal areas. Figure 9c shows the spatial distribution of CD values on the annual scale. All CD values are between 1.73 and 2.99, and the spatial pattern is similar with the spatial pattern on the seasonal scale. As we all know, the eastern coastal areas, especially Shanghai, Suzhou, and Hangzhou, are densely populated and have high levels of urbanization, and the CD value of this area is relatively low, while the areas located in the northwest of the YRD, such as Bozhou, Xuzhou, and Fuyang, the large outflow of people results in a relatively small population in these areas, and the urbanization level is relatively low, and the CD value in this area is relatively high. It can be seen that the population density and the urbanization level are related to CD.

In general, on different temporal scales, the high values of CD are mainly distributed in the sparse areas of population and urban, while the low values of CD are mainly distributed in the dense areas of population and urban.

3.2.3. The Influences of Driving Factors and Their Interactions on Temperature Slope

From the above results, we can see that the spatial distribution of TS is different, and what is the reason for this result? In order to answer this question, we choose some driving factors (AT, NDVI, UD, GDP, and NL) that affect the temperature to explore the reasons of this phenomenon by using the Geogdetector method.

The factor detector is used to detect whether the driving factors affect TS and the size of their influences. In addition, the greater the value of q, the greater the influence of this factor on TS. Table 2 shows the result of the factor detector. On the whole, the influence, in order of size, of each factor is: UD (0.323) > GDP (0.234) > NL (0.218) > NDVI (0.118) > AT (0.047). In addition, all driving factors pass the significant test, which means that these five factors have significant effects on TS. In addition, we can see that the contribution rate of socioeconomic factors (UD, GDP, NL) is greater than natural factors (NDVI, AT).

Table 2. The result of factor detectors.

	GDP	AT	NL	UD	NDVI
q statistic	0.234	0.047	0.218	0.323	0.118
p-value	0.000	0.000	0.000	0.000	0.000

Note: GDP represents the gross domestic product; AT represents the altitude; NL represents the night light; UD represents the urban density; NDVI represents the normalized difference vegetation index.

Whether the factor has a significant difference in the spatial distribution affecting the TS is achieved by an ecological detector. A test with a significance level of 0.05 indicates that the two factors are different influencing the distribution of TS; otherwise, there is no significant difference. The result of an ecological detector is shown in Table 3.

Table 3. The result of an ecological detector.

		Socio-Economic Factors			Natural Factors	
		GDP	UD	NL	AT	NDVI
Socio-economic factors	GDP	-	-	-	-	-
	UD	Y	-	-	-	-
	NL	N	Y	-	-	-
Natural factors	AT	N	Y	Y	-	-
	NDVI	N	N	N	Y	-

Note: Y indicates that the two factors have significant differences in the spatial distribution of temperature slopes, N indicates no significant difference, and the confidence is 95%. And GDP represents the gross domestic product; UD represents the urban density; NL represents the night light; AT represents the altitude; NDVI represents the normalized difference vegetation index.

The result shows that there is a significant difference between UD and GDP; there is no significant difference between AT, NL, NDVI, and GDP, indicating that the effects of AT, NL, NDVI, and GDP on the spatial distribution of TS are similar. In addition, there is a significant difference between UD and NL, AT, and there is no significant difference between UD and NDVI. Similarly, there is a significant difference between NL and AT, while NL is not significantly different from NDVI. For AT and NDVI, there is also a significant difference between them. We can also conclude that the influences of various driving factors on the TS are different.

Table 2 indicates that the contribution rate of each factor alone to the TS is different. Thus, there is an interaction between them, and, if so, what is the interaction result? In order to answer this question, we give the results of interaction detector as Table 4.

Table 4. The result of interaction detectors.

		Socio-Economic Factors GDP			Natural Factors	
		GDP	UD	NL	AT	NDVI
Socio-Economic	GDP	0.234	-	-	-	-
Factors GDP	UD	0.464 [#]	0.323	-	-	-
Natural Factors	NL	0.391 [#]	0.420 [#]	0.218	-	-
	AT	0.290 [*]	0.365 [#]	0.235 [#]	0.047	-
	NDVI	0.314 [#]	0.393 [#]	0.262 [#]	0.146 [#]	0.118

Note: # indicates that the interaction is a bi-enhancement, i.e., q ($X1 \cap X2$) > Max($q(X1)$, $q(X2)$); * indicates that the interaction is a nonlinear enhancement, i.e., q ($X1 \cap X2$) > $q(X1)$ + $q(X2)$. And GDP represents the gross domestic product; UD represents the urban density; NL represents the night light; AT represents the altitude; NDVI represents the normalized difference vegetation index.

Table 4 shows that only AT and GDP have a nonlinear enhancement effect (q (GDP$\cap$ AT) > q(GDP) + q(AT)) on TS, and the interactions between remaining driving factors have the bi-enhancement effect on TS. It shows that the effect of interaction of any two factors is greater than the effect of a single factor. Among them, the interaction effect between GDP and UD (q (GDP $\cap$ UD) = 0.464) is the largest, and the interaction effect between UD and NL (q (UD$\cap$ NL) = 0.420) is second, followed by the interaction effect between UD and NDVI (q (UD$\cap$NDVI) = 0.393) and the interaction effect between GDP and NL (q (GDP$\cap$ NL) = 0.391), while the interaction effect between AT and NDVI (q (AT$\cap$NDVI) = 0.146) is the smallest. In general, the interaction effect between socioeconomic factors is the largest, the interaction effect between socioeconomic factors and natural factors is second, followed by the interaction effect between natural factors.

4. Discussion

In this study, we found that the temperature rose 1.53 °C with an average rising rate of 0.465 °C/10 years during the study period, which was higher than the global average rate. The result was consistent with previous studies [37–39]. It confirms the regional differences in climate change. In addition, it means that the temperature was not only affected by global warming, but also affected by various driving factors within the region. In addition, the temperature rose quickly in the dense areas of population and urban, and the temperature rose slowly in the sparse areas of population and urban. It reflected the urban-rural differences in temperature distribution from the side.

The climate system was an open system with external forcing and nonlinear dissipation [40], and fractal theory was one of the effective methods to quantitatively describe the nonlinear evolution process of climate and its self-similar structural features. Numerous studies [41–45] had shown that fractal analysis could calculate its fractal dimension from a seemingly chaotic climate sequence, confirming the fractal information of the climate system. Temperature was an element of the climate system and also had nonlinear characteristics. Especially in the YRD, the temperature was more complicated due to the influence of human activities. By calculating the CDs on the daily, seasonal, and annual scales of the YRD, we confirmed that the temperature in the YRD was a chaotic dynamic system with nonlinear characteristics. We found that temperature on the annual scale was more complicated than on the daily scale in the YRD. It was because the annual average temperature was the average of the daily temperature, which was the macroscopic performance of the daily temperature and influenced by many factors [46,47], so it showed greater complexity on the whole. Xu et al. [17] found that the temperature process on the daily scale was more complicated than the temperature process on the annual scale in Xinjiang. It was contrary to the YRD, indicating the complexity of the temperature process had regional differences. In the spatial distribution, whether in the daily, seasonal, or annual scale, the high CD values were mainly distributed in the sparse areas of population and urban, while the low CD values were mainly distributed in the dense areas of population and urban. In the dense areas of population and urban, due to the density of cities and people, industrial and urbanization were developing rapidly, and the temperature was mainly affected by the rapid development of cities, showing an upward trend [38]. While in the sparse areas of population and urban, the temperature changes were mainly affected by natural factors and socioeconomic factors together, so the temperature changes more complicated.

The effects of five driving factors on the TS were quantitatively investigated by using the Geogdetector method. UD was the most important factor affecting TS. From Figures 2c and 8, we can see that the spatial distribution of UD was similar to the spatial distribution of TS, that is, decreasing toward the periphery with Shanghai as the center. The UD reflected the intensity of the city. In Shanghai and its surrounding areas, cities were dense and urbanization was high. One of the most striking features of this was that the impervious surface of the city increased rapidly [48,49]. The impervious surface of the city had strong heat storage, poor water storage capacity, and hindering airflow transmission, which seriously affected the city's surface hydrological cycle [50], energy distribution [49] and urban microclimate [51], resulting in an urban heat island effect, which causes the temperature in dense areas of urban to rise quickly. The impact of urban impervious surface on surface temperature had been verified in different regions and a certain consensus had been reached [52–55]. The contribution rate of GDP to TS was second. From Figures 2 and 6, we can see that the spatial distribution of GDP was similar to the spatial distribution of TS. The development of GDP inevitably consumed a large amount of energy, which would emit a large amount of greenhouse gases, resulting in a quick increase in temperature. The GDP in Shanghai and its surrounding areas was increasing rapidly, so the TS in this area was high. The contribution of NL was similar to the contribution rate of GDP, and the spatial distribution of NL was similar to the TS. NL reflected the level of GDP and energy consumption from the side [56,57], so it had a high contribution rate to TS. The NDVI was the smallest in Shanghai and its surrounding area, while the TS was largest in this area, which meant that the vegetation coverage rate played an important role in suppressing the increase of temperature, but it was not enough only to rely on the vegetation coverage rate. Most of the YRD was plain and the fluctuation of terrain was small, so the contribution rate of AT to the TS was small and can be ignored. Each driving factor had an effect on TS; they did not work alone, but different driving factors interacted with each other and had an enhanced influence on temperature.

Our main purpose is to explore which factors or interactions between factors contribute the most to temperature. The paper only analyzed the complexity of the temperature process and the contribution rates of driving factors to temperature, but the mechanism behind it remains to be studied further. In addition, through the analysis of the driving factors, some policy opinions to mitigate the temperature rise need to be proposed in the next study.

5. Conclusions

The study first analyzed the spatiotemporal variations of temperature in the YRD during the period of 1980–2012 by using the trend analysis method; then, we investigated the spatiotemporal complexity of temperature on different time scales by using the correlation dimension; finally, the effects of driving factors and their interactions on TS in the YRD during the period of 1980–2012 was analyzed by using the Geogdetector method. Summarizing this study, the main conclusions are as follows:

1. The temperature was increasing during the period of 1980–2012, and it rose by 1.53 °C from 1980 to 2012; in addition, among the dense areas of population and urban, the temperature rose quickly, while the temperature in the sparse areas of population and urban rose slowly.
2. In the temporal, the temperature process was more complicated with the increase of temporal scale; in the spatial distribution, whether it is the daily time scale, the seasonal time scale, or

the annual time scale, the temperature process was more complicated in the sparse areas of population and urban than the dense areas of population and urban.

3. Socioeconomic factors were the main factors affecting climate change in the YRD, and the contribution rate of urban density is the largest among the contribution rates of single factors. In addition, the interactions between various driving factors had an enhanced effect on regional climate change. In addition, the interaction between economic activity and urban density had the largest influence on temperature.

Author Contributions: C.Z. designed, carried out the analysis, and wrote the manuscript. N.Z. revised the paper and refined the results, conclusion, and abstract. D.Y. discussed the results. N.Z. edited the figures. J.X. revised the paper and gave the commentary. All authors have read and agreed to the published version of the manuscript.

Funding: This research was funded by the Scientific and Technological Innovation Programs of Higher Education Institutions in Shanxi, Grant No. 2019L0477 and the Humanities and Social Sciences Foundation of the Ministry of Education of China, Grant No. 19YJA890006.

Acknowledgments: The authors are grateful to the Resource and Environmental Science Data Center (http://www. resdc.cn) of the Chinese Academy of Sciences and the China Meteorological Data Sharing Service System (http: //cdc.cma.gov.cn/) for providing data. The authors appreciate the insightful comments of anonymous reviewers.

Conflicts of Interest: The authors declare no conflict of interest.

References

1. Wang, B.; Zhang, M.; Wei, S.; Wang, S.; Li, S.; Ma, Q.; Li, X.; Pan, S. Changes in extreme events of temperature and precipitation over Xinjiang, northwest China, during 1960–2009. *Quatern. Int.* **2013**, *298*, 141–151. [CrossRef]
2. Horton, D.E.; Johnson, N.C.; Singh, D.; Swain, D.L.; Rajaratnam, B.; Diffenbaugh, N.S. Contribution of changes in atmospheric circulation patterns to extreme temperature trends. *Nature* **2015**, *522*, 465–469. [CrossRef] [PubMed]
3. Fu, P.; Weng, Q. A time series analysis of urbanization induced land use and land cover change and its impact on land surface temperature with Landsat imagery. *Remote Sens. Environ.* **2016**, *175*, 205–214. [CrossRef]
4. de Barros Soares, D.; Lee, H.; Loikith, P.C.; Barkhordarian, A.; Mechoso, C.R. Can significant trends be detected in surface air temperature and precipitation over South America in recent decades? *Int. J. Climatol.* **2017**, *37*, 1483–1493. [CrossRef]
5. Liuzzo, L.; Bono, E.; Sammartano, V.; Frenia, G. Long-term temperature changes in Sicily, Southern Italy. *Atmos. Res.* **2017**, *198*, 44–55. [CrossRef]
6. Zhu, N.; Xu, J.; Li, W.; Li, K.; Zhou, C. A Comprehensive Approach to Assess the Hydrological Drought of Inland River Basin in Northwest China. *Atmosphere* **2018**, *9*, 370. [CrossRef]
7. Ullah, S.; You, Q.; Ali, A.; Ullah, W.; Lan, M.A.; Zhang, Y.; Xie, W.; Xie, X. Observed changes in maximum and minimum temperatures over China- Pakistan economic corridor during 1980–2016. *Atmos. Res.* **2019**, *216*, 37–51. [CrossRef]
8. Sharma, C.S.; Panda, S.N.; Pradhan, R.P.; Singh, A.; Kawamura, A. Precipitation and temperature changes in eastern India by multiple trend detection methods. *Atmos. Res.* **2016**, *180*, 211–225. [CrossRef]
9. Salman, S.A.; Shahid, S.; Ismaila, T.; Ahmed, K.; Wang, X.J. Selection of climate models for projection of spatiotemporal changes in temperature of Iraq with uncertainties. *Atmos. Res.* **2018**, *213*, 509–522. [CrossRef]
10. Kenawy, A.E.; López-Moreno, J.I.; Vicente-Serrano, S. Trend and variability of surface air temperature in northeastern Spain (1920–2006): Linkage to atmospheric circulation. *Atmos. Res.* **2012**, *106*, 159–180. [CrossRef]
11. Iqba, M.A.; Penas, A.; Cano-Ortiz, A.; Kersebaum, K.C.; Herrero, L.; del Rio, L. Analysis of recent changes in maximum and minimum temperatures in Pakistan. *Atmos. Res.* **2016**, *168*, 234–249. [CrossRef]
12. Baliunas, S.; Frick, P.; Sokoloff, D.; Soon, W. Time scales and trends in the Central England Temperature data (1659–1990): A wavelet analysis. *Geophys. Res. Lett.* **1997**, *24*, 1351–1354. [CrossRef]
13. Bolzan, M.J.A.; Vieira, P.C. Wavelet Analysis of the Wind Velocity and Temperature Variability in the Amazon Forest. *Braz. J. Phys.* **2006**, *36*, 1217–1222. [CrossRef]

14. Wu, Z.; Huang, N.E.; Wallace, J.M.; Smoilak, B.V.; Chen, X. On the time-varying trend in global-mean surface temperature. *Clim. Dyn.* **2011**, *37*, 759–773. [CrossRef]

15. Macias, D.; Stips, A.; Garcia-Gorriz, E. Application of the Singular Spectrum Analysis Technique to Study the Recent Hiatus on the Global Surface Temperature Record. *PLoS ONE* **2014**, *9*, e107222. [CrossRef]

16. Dawood, M. Spatio-statistical analysis of temperature fluctuation using Mann-Kendall and Sen's slope approach. *Clim. Dyn.* **2017**, *48*, 783–797. [CrossRef]

17. Xu, J.; Chen, Y.; Li, W.; Liu, Z.; Wei, C.; Tang, J. Understanding the complexity of temperature dynamics in Xinjiang, China, from multitemporal scale and spatial perspectives. *Sci. World J.* **2013**, *2013*, 259248. [CrossRef]

18. Najafi, M.R.; Zwiers, F.W.; Gillett, N.P. Attribution of Arctic temperature change to greenhouse-gas and aerosol influences. *Nat. Clim. Chang.* **2015**, *5*, 246–249. [CrossRef]

19. Chen, Y.; Chiu, H.; Su, Y.; Wu, Y.C.; Cheng, K.S. Does urbanization increase diurnal land surface temperature variation? Evidence and implications. *Landsc. Urban Plan* **2017**, *157*, 247–258. [CrossRef]

20. Shi, J.; Cui, L.; Ma, Y.; Du, H.; Wen, K. Trends in temperature extremes and their association with circulation patterns in China during 1961–2015. *Atmos. Res.* **2018**, *212*, 259–272. [CrossRef]

21. Liang, L.; Engling, G.; Zhang, X.; Sun, J.; Zhang, Y.; Wu, W.; Liu, C.; Zhang, G.; Liu, X.; Ma, Q. Chemical characteristics of PM2.5 during summer at a background site of the Yangtze River Delta in China. *Atmos. Res.* **2017**, *198*, 163–172. [CrossRef]

22. Zhang, Q.; Gemmer, M.; Chen, J. Climate changes and flood/drought risk in the Yangtze Delta, China, during the past millennium. *Quatern. Int.* **2008**, *176*, 62–69. [CrossRef]

23. Chu, W.; Qiu, S.; Xu, J. Temperature Change of Shanghai and Its Response to Global Warming and Urbanization. *Atmosphere* **2016**, *7*, 114. [CrossRef]

24. Kawashima, S. Effect of vegetation on surface temperature in urban and suburban areas in winter. *Energy Build.* **1990**, *15*, 465–469. [CrossRef]

25. Gall, K.P.; Elvidge, C.D.; Yang, L.; Reed, B.C. Trends in night-time city lights and vegetation indices associated with urbanization within the conterminous USA. *Int. J. Remote Sens.* **2004**, *25*, 2003–2007. [CrossRef]

26. Zhang, N.; Gao, Z.; Wang, X.; Chen, Y. Modeling the impact of urbanization on the local and regional climate in Yangtze River Delta, China. *Theor. Appl. Climatol.* **2010**, *102*, 331–342. [CrossRef]

27. Hess, A.; Iyer, H.; Malm, W. Linear trend analysis: A comparison of methods. *Atmos. Environ.* **2001**, *35*, 5211–5222. [CrossRef]

28. Oliver, M.A.; Webster, R. Kriging: A method of interpolation for geographical information systems. *Int. J. Geogr. Inf. Syst.* **1990**, *4*, 313–332. [CrossRef]

29. Grassberger, P.; Procaccia, I. Measuring the strangeness of strange attractors. *Phys. D* **1983**, *9*, 189–208. [CrossRef]

30. Sivakumar, B. Nonlinear determinism in river flow: Prediction as a possible indicator. *Earth Surf. Proc. Land.* **2006**, *32*, 969–979. [CrossRef]

31. Ling, H.; Xu, H.; Fu, J.; Zhang, Q.; Xu, X. Analysis of temporal-spatial variation characteristics of extreme air temperature in Xinjiang, China. *Quatern. Int.* **2012**, *282*, 14–26. [CrossRef]

32. Wang, J.; Li, X.H.; Christakos, G.; Liao, Y.; Zhang, T.; Gu, X.; Zheng, X. Geographical Detectors-Based Health Risk Assessment and its Application in the Neural Tube Defects Study of the Heshun Region, China. *Int. J. Geogr. Inf. Sci.* **2010**, *24*, 107–127. [CrossRef]

33. Yang, D.; Wang, X.; Xu, J.; Xu, C.; Lu, D.; Ye, C.; Wang, Z.; Bai, L. Quantifying the influence of natural and socioeconomic factors and their interactive impact on PM 2.5 pollution in China. *Environ. Pollut.* **2018**, *241*, 475–483. [CrossRef]

34. Wang, J.; Hu, Y. Environmental health risk detection with GeogDetector. *Environ. Model. Softw.* **2012**, *33*, 114–115. [CrossRef]

35. Qi, L.; Wang, Y. Changes in the observed trends in extreme temperatures over China around 1990. *J. Clim.* **2012**, *25*, 5208–5222. [CrossRef]

36. Grassberger, P.; Procaccia, I. Characterization of Strange Attractors. *Phys. Rev. Lett.* **1983**, *50*, 346. [CrossRef]

37. Yang, B.; Braeuning, A.; Johnson, R.K.; Shi, Y.F. General characteristics of temperature variation in China during the last two millennia. *Geophys. Res. Lett.* **2002**, *29*, 31–38. [CrossRef]

38. Du, Y.; Xie, Z.; Zeng, Y.; Shi, Y.; Wu, J. Impact of urban expansion on regional temperature change in the Yangtze River Delta. *J. Geogr. Sci.* **2007**, *17*, 387–398. [CrossRef]

39. Yan, H.; Zhang, J.; Hou, Y.; He, Y. Estimation of air temperature from MODIS data in east China. *Int. J. Remote Sens.* **2009**, *30*, 6261–6275. [CrossRef]

40. Rial, J.; Coauthors, A. Nonlinearities, feedbacks and critical thresholds within the Earth's climate system. *Clim. Chang.* **2004**, *65*, 11–38. [CrossRef]

41. Olsson, J.; Niemczynowicz, J.; Berndtssonet, R. Fractal Analysis of High-Resolution Rainfall Time Series. *J. Geophys. Res. Atmos.* **1993**, *98*, 23265–23274. [CrossRef]

42. Bodri, L. Fractal analysis of climatic data: Mean annual temperature records in Hungary. *Theor. Appl. Climatol.* **1994**, *49*, 53–57. [CrossRef]

43. Radziejewski, M.; Kundzewicz, Z.W. Fractal analysis of flow of the river Warta. *J. Hydrol.* **1997**, *200*, 80–294. [CrossRef]

44. Oñate Rubalcaba, J.J. Fractal analysis of climatic data: Annual precipitation records in Spain. *Theor. Appl. Climatol.* **1997**, *56*, 83–87. [CrossRef]

45. Rangarajan, G.; Sant, D.A. Fractal dimensional analysis of Indian climatic dynamics. *Chaos Solitons Fract.* **2004**, *19*, 285–291. [CrossRef]

46. Webster, P.J.; Yang, S. Monsoon and ENSO: Selectively interactive systems. *Q. J. R. Meteorol. Soc.* **1992**, *118*, 877–926. [CrossRef]

47. Zhang, K.; Wang, R.; Shen, C.; Da, L. Temporal and spatial characteristics of the urban heat island during rapid urbanization in Shanghai, China. *Environ. Monit. Assess.* **2020**, *169*, 101–112. [CrossRef]

48. Wu, C.; Murray, A.T. Estimating impervious surface distribution by spectral mixture analysis. *Remote Sens. Environ.* **2003**, *84*, 493–505. [CrossRef]

49. Yuan, F.; Bauer, M.E. Comparison of impervious surface area and normalized difference vegetation index as indicators of surface urban heat island effects in Landsat imagery. *Remote Sens. Environ.* **2007**, *106*, 375–386. [CrossRef]

50. Brun, S.E.; Band, L.E. Simulating runoff behavior in an urbanizing watershed. *Comput. Environ. Urban Syst.* **2000**, *24*, 5–22. [CrossRef]

51. Xu, H. Analysis of Impervious Surface and its Impact on Urban Heat Environment using the Normalized Difference Impervious Surface Index (NDISI). *Photogramm. Eng. Remote Sens.* **2010**, *76*, 557–565. [CrossRef]

52. Xiao, R.B.; Ouyang, Z.Y.; Zhang, H.; Li, W.; Schienke, E.W.; Wang, X. Spatial pattern of impervious surfaces and their impacts on land surface temperature in Beijing, China. *J. Environ. Sci.* **2007**, *19*, 250–256. [CrossRef]

53. Weng, Q.; Lu, D. A sub-pixel analysis of urbanization effect on land surface temperature and its interplay with impervious surface and vegetation coverage in Indianapolis, United States. *Int. J. Appl. Earth Obs. Geoinf.* **2008**, *10*, 68–83. [CrossRef]

54. Zhang, Y.; Odeh, I.O.A.; Han, C. Bi-temporal characterization of land surface temperature in relation to impervious surface area, NDVI and NDBI, using a sub-pixel image analysis. *Int. J. Appl. Earth Obs.* **2009**, *11*, 256–264. [CrossRef]

55. Nie, Q.; Xu, J. Understanding the effects of the impervious surfaces pattern on land surface temperature in an urban area. *Front. Earth Sci.* **2015**, *9*, 276–285. [CrossRef]

56. Doll, C.N.H.; Muller, J.; Morleyet, J.G. Mapping regional economic activity from night-time light satellite imagery. *Ecol. Econ.* **2006**, *57*, 75–92. [CrossRef]

57. Elvidge, C.D.; Baugh, K.E.; Anderson, S.; Sutton, P.C.; Ghosh, T. The Night Light Development Index (NLDI): A spatially explicit measure of human development from satellite data. *Soc. Geogr.* **2012**, *7*, 23–35. [CrossRef]

Article

Coupling of Soil Moisture and Air Temperature from Multiyear Data During 1980–2013 over China

Qing Yuan, Guojie Wang *, Chenxia Zhu, Dan Lou, Daniel Fiifi Tawia Hagan, Xiaowen Ma and Mingyue Zhan

Collaborative Innovation Center on Forecast and Evaluation of Meteorological Disasters, School of Geographical Sciences, Nanjing University of Information Science & Technology, Nanjing 210044, China; qyuan@nuist.edu.cn (Q.Y.); zhuchenxia@nuist.edu.cn (C.Z.); loudan711@163.com (D.L.); dhagan@yeah.net (D.F.T.H.); 17853462199@163.com (X.M.); zhanmingyue0614@163.com (M.Z.)
* Correspondence: gwang@nuist.edu.cn; Tel.: +86-25-58731418

Received: 28 November 2019; Accepted: 24 December 2019; Published: 26 December 2019

Abstract: Soil moisture is an important parameter in land surface processes, which can control the surface energy and water budgets and thus affect the air temperature. Studying the coupling between soil moisture and air temperature is of vital importance for forecasting climate change. This study evaluates this coupling over China from 1980–2013 by using an energy-based diagnostic method, which represents the momentum, heat, and water conservation equations in the atmosphere, while the contributions of soil moisture are treated as external forcing. The results showed that the soil moisture–temperature coupling is strongest in the transitional climate zones between wet and dry climates, which here includes Northeast China and part of the Tibetan Plateau from a viewpoint of annual average. Furthermore, the soil moisture–temperature coupling was found to be stronger in spring than in the other seasons over China, and over different typical climatic zones, it also varied greatly in different seasons. We conducted two case studies (the heatwaves of 2013 in Southeast China and 2009 in North China) to understand the impact of soil moisture–temperature coupling during heatwaves. The results indicated that over areas with soil moisture deficit and temperature anomalies, the coupling strength intensified. This suggests that soil moisture deficits could lead to enhanced heat anomalies, and thus, result in enhanced soil moisture coupling with temperature. This demonstrates the importance of soil moisture and the need to thoroughly study it and its role within the land–atmosphere interaction and the climate on the whole.

Keywords: soil moisture–temperature coupling; heatwaves; multiple time scales

1. Introduction

The summer of 2013 was unprecedentedly hot in Eastern China, causing substantial societal and economic impacts [1]. Such a phenomenon has drawn widespread concerns, and the physical mechanism behind such heatwave is gradually being discovered [2–4]. The changes of large-scale atmospheric circulations may be the main cause of temperature anomalies, and small-scale physical processes of local energy balance such as soil moisture–atmosphere coupling could also make a contribution to them [5]. Many studies have shown that soil moisture anomalies play an important role in soil moisture–temperature coupling [6,7], as it could control the energy budget by the partitioning of latent heat flux and sensible heat flux, further impacting the air temperature [8,9]. When soil moisture decreases, less water can be used for evapotranspiration, resulting in a decrease of latent heat flux [8,10]. Based on the energy balance, the decline of latent heat flux causes an increase of the sensible heat flux, thus enhancing the air temperature. These conditions indicate a negative feedback between soil moisture and air temperature: Soil moisture deficit results in the rise of air temperature [11–13].

Nowadays, various studies have shown that dry soil moisture conditions can have a substantial influence on the severity of heat waves and drought through the coupling between soil moisture and atmosphere [10,11]. As shown, soil moisture–temperature coupling helps to explain heat waves in summer climate [14,15]. Modeling experiments have focused on identifying the strong coupling regions, and how these regions are influenced by the changing climate [14,16]. The Global Land–Atmosphere Coupling (GLACE) project indicated that the strongest coupling regions (hot spots) of soil moisture–temperature are between the transitional regions of wet and dry climates [16,17]. In addition, some numerical experiments have been devoted to studying the soil moisture–temperature coupling at regional scales [18–20] finding that soil moisture anomalies impact air temperature during summer mainly in areas like Northern China [21–23].

At present, many studies have tried to use different metrics for assessing land–atmosphere coupling strengths. Koster et al. [16] proposed to use correlation between evapotranspiration and temperature, and found results agreeing with other metrics, for example, those based on the correlation between evapotranspiration and radiation. Gallego-Elvira et al. [24] used the dependence of the surface heating rate of different cover types on the previous precipitation during drought to identify different evaporative regions. Dirmeyer [25] devoted an index of surface flux sensitivity to soil moisture variability and applied it to global atmospheric reanalysis datasets. However, most studies assessing the coupling of soil moisture to climate were based primarily on summer (June–July–August, JJA) [9,14], but there is a lack of understanding on its seasonal changes; therefore, it is important to devote more attention to studying the coupling within the other seasons as well. Moreover, the studies of the soil moisture–temperature coupling were mostly based on modeling experiments [10,26], which show a large difference in the regions and strengths of land–atmosphere coupling [8]. However, the limited ground measurements of soil moisture cannot meet the research needs of land–atmosphere coupling on regional scales. Given the limitation of ground measurements, satellite data could be used in soil moisture–temperature coupling studies from an observational point of view, and the recent development of soil moisture and evapotranspiration products from remote sensing technology provides the possibility for such studies.

A wide variety of datasets makes it possible for us to study soil moisture–temperature coupling from an observational perspective [24,27,28]. This study utilized one of these satellite-based datasets with a coupling diagnostic to show the spatial distribution and interannual variation of strong coupling regimes between soil moisture and temperature over China in different seasons. This diagnostic focused on two different timescales to fill the gap between extremes and climatological studies of soil moisture–temperature coupling. Different season coupling hot spots of China are illustrated in the following sections. Subsequently, we explored the role of soil moisture during the 2013 heatwave in Southeast China and the 2009 event in North China.

Furthermore, this study depicts not only soil moisture–temperature coupling in long-term variations, but also soil moisture–temperature coupling during the heat wave events and the related heating processes. It may help to point us toward better understanding the underlying processes of soil moisture–temperature coupling, and thus improve the prediction skills of heat waves [14,29].

2. Materials and Methods

2.1. Study Area

China has a complex climate due to its topography [30]. Mainland China is generally divided into three different climatic zones: Arid region with annual precipitation below 200 mm, humid region with annual precipitation more than 800 mm, and transitional region with annual precipitation from 200 to 800 mm [31]. The arid region mainly includes Xinjiang province and western Inner Mongolia Plateau. The humid region is mainly South China. The transitional region mainly includes North China Plain, Northeast Plain, and part of Tibetan Plateau, where there is more rain in summer but less in winter. Known as the "The Third Pole", the Tibetan Plateau is the highest and most unique geographical unit on earth [32]. It is one of the most sensitive regions to global climate change and its hydrological processes are quite different from that of other regions of China [33,34]; thus, it is regarded as a separate typical region in this study. The humid, transitional, and dry regions, as well as the Tibetan Plateau, are shown in Figure 1.

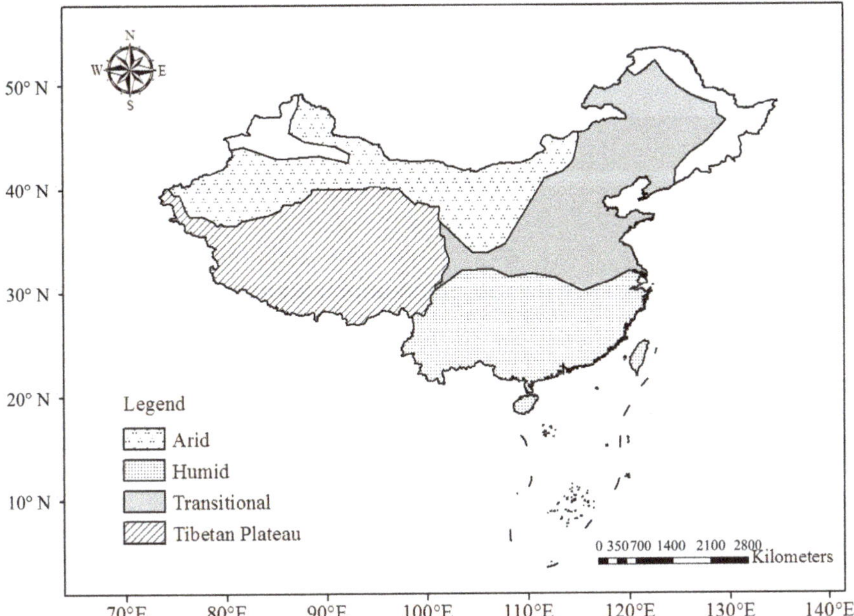

Figure 1. The typical climate regions of China (arid, humid, transitional, the Tibetan Plateau).

2.2. Data Sources

The evapotranspiration (ET) and potential evapotranspiration (PET) data from the GLEAM v3.0a (Global Land Evaporation Amsterdam Model) product were used in this study, which span the period from 1980 to 2015 with a spatial resolution of 0.25°. The GLEAM model is a simplified land model, which is fully dedicated to estimating the terrestrial evaporation and root zone soil moisture based on satellite data [35]. It comprises a set of algorithms using multiyear satellite observations to estimate the components of terrestrial ET. The PET is calculated within the Priestley–Taylor equation via the observations of net radiation and near-surface air temperature [36]. The 2-m air temperature and the top layer (0–7 cm) volumetric soil moisture from the ERA-Interim reanalysis data were used [37].

2.3. Methods

This study used a diagnostic method to estimate the long-term soil moisture–temperature coupling over China in different seasons, which is based on two energy balances of ET and PET. The partitioning of the land energy is expressed in Equation (1), where R_n refers to the surface net radiation, G means the ground heat flux, and λ is the latent heat of vaporization, which can be captured from near-surface air temperature; the ground heat flux is negligible in this study [36,38].

$$R_n - G = \lambda E + H,\tag{1}$$

When the annual time series of E, E_p, R_n, and near-surface temperature (T) are available, the diagnostic method could be used to estimate the long-term soil moisture–temperature coupling [17], where the metric (Π) can be calculated as:

$$\Pi = \rho(H,T) - \rho(H_p, T), \tag{2}$$

$$H = R_n - \lambda E, \tag{3}$$

$$H_p = R_n - \lambda E_p, \tag{4}$$

where ρ means Pearson's correlation coefficient, and H is the sensible heat flux. Using Π, we can derive an indicator of the long-term soil moisture–temperature coupling, which can be considered as a multi-year average. When considering the σ_T, σ_H, and σ_{H_p} (the standard deviations of T, H, and H_p), Equation (1) could also be expressed in another form as covariances:

$$\Pi = \frac{1}{\sigma_T}\left(\frac{cov(H,T)}{\sigma_H} - \frac{cov(H_p,T)}{\sigma_{H_p}}\right), \tag{5}$$

To understand the related heating processes between land and atmosphere during the heatwave events, we used a different diagnostic method based on daily data to derive the soil moisture–temperature coupling at daily scale [8], and the metric (π) is defined as:

$$\pi_i = \frac{T_i - \overline{T}}{\sigma_T}\left(\frac{H_i - \overline{H}}{\sigma_H} - \frac{H_{p,i} - \overline{H_p}}{\sigma_{H_p}}\right), \tag{6}$$

where $\overline{T}$, $\overline{H}$, and $\overline{H_p}$ indicate the averages of T, H, and H_p over a long term. It can be simplified as:

$$\pi = T' \times e', \tag{7}$$

$$e' = (R_n - \lambda E)' - (R_n - \lambda E_p)', \tag{8}$$

where T' represents the anomalies of T, and e' is equal to $H' - H_p'$ and indicates the contribution of soil moisture deficit to sensible heat flux. When there is sufficient soil moisture for the atmospheric demand, this energy term will be zero, and it may increase under arid condition. Only if the potential influence of soil moisture on temperature is accompanied by a large anomalous value of temperature is the local energy balance likely to control air temperature [11].

The Π and the π are two coupling metrics at different time scales; the former includes the long-term record in terms of correlation coefficients to evaluate long-term climatology, while the latter expresses anomalies of one day in terms of standard deviations to evaluate daily extreme. When Π and π values are greater than zero, then the higher the value the stronger soil moisture–temperature coupling. If values are less than or equal to zero, there is no coupling [17].

3. Results

3.1. Long-Term Soil Moisture–Temperature Coupling

To know the spatial distribution of strong coupling between soil moisture and temperature in China, we first calculated the metric (Π) by using ET (evapotranspiration) and PET (potential evapotranspiration) from the GLEAM data, and T (2-meter temperature) and Rn (surface net radiation) from the ERA-Interim reanalysis data over the period 1980–2013. Figure 2 illustrates the derived long-term soil moisture–temperature coupling annually and in different seasons.

Figure 2a shows the Π values derived from the annual data. The coupling strength appears to be highest in Northeast China and part of the Tibetan Plateau, where the climate is neither too wet nor too dry, showing that soil moisture has the strongest impact on temperature over the annual average in these regions. The results are consistent with previous studies which have suggested that such hot spots of soil moisture–temperature coupling occur most in transitional regions between wet and dry climate [9,39,40]. In spring, there appears to be strong soil moisture–temperature coupling

(Figure 2b) in large areas of China, including Northeast, North, Northwest China and the Tibetan Plateau, and Yunnan province as well.

In summer, the coupling strengths are largely reduced compared to those in spring. There are only coupling signals in North China, especially Inner Mongolia. In autumn, the strongest coupling signals appear in Northeast China and the northern part of the Tibetan Plateau, although these are also significantly weakened compared to those in spring. It is not surprising to find strong signals in Northeast China, where it is neither too wet nor too dry in summer. In the cold season of winter, quite limited coupling signals are found, indicating that the soil moisture generally has quite limited impact on temperature across China.

Over the whole region, as shown in Figure 2, soil moisture–temperature coupling is relatively stronger in spring, followed by summer and autumn, and rather insignificant in winter. The seasonality of the soil moisture–temperature coupling strength has a distinct regional variation. Over Northern China, which is mainly a arid/semi-arid region, the contribution of soil moisture to evapotranspiration is mainly limited by water. Here, the results in Figure 2 depict strong coupling, which suggests a stronger impact of soil moisture anomalies on temperature. Given that Southern China is mainly a humid/semi-humid region, the comparatively weaker soil moisture–temperature coupling (as seen in Figure 2) demonstrates that the region is mostly dominated by energy-limited conditions [8,13]. The north of China is here identified as a hot spot in spring, mainly because dry conditions (water-limited) persist throughout the year. In spring, a sufficient energy supply for ET, due to melting ice and snow, increases the amount of water in these regions, as well as causes more evaporation and, thus, increased coupling strength to the atmosphere. This may explain why coupling strength is higher in spring than in summer in North China [41]. The seasonal transition from winter to spring affects the soil moisture thawing and radiation budget over the Tibetan Plateau, which results in more heat transfer into the atmosphere. The heat energy transferred to the atmosphere is used to warm the air, thus showing strong soil moisture–temperature coupling [42]. In Yunnan province, the water supply is not sustainable because of the special climatic conditions, thus the soil is drier during the spring, which leads to a strong soil moisture–temperature coupling [43].

Figure 3 shows the soil moisture–temperature coupling strengths in different seasons based on the four climatic regions. Obviously, the coupling strengths vary greatly in different regions and different seasons. Except for the humid region, the coupling strengths of the other three climatic regions are all the strongest in spring and the weakest in winter. From the perspective of different seasons, the strength differences between the arid region, the transitional region, and the Tibetan Plateau region are relatively small. The seasonal transition from winter to spring influences soil moisture thawing and radiation budgets, with more heat energy being transferred into atmosphere [42,43]. For the typical climate regions divided according to Figure 1, the arid and transitional regions and the Tibetan Plateau zone have analogous soil moisture and atmosphere conditions, thus they are very similar, particularly in spring. However, coupling strengths in the humid region appear to be much smaller than the other regions in all seasons. The transitional region appears to have the strongest coupling strength in terms of the annual average, followed by the arid region and the Tibetan Plateau. The soil moisture–temperature coupling in spring appears to be much stronger than in the other seasons, which is particularly significant in the arid and transitional regions and the Tibetan Plateau. In the cold winter, the soil moisture–temperature coupling strengths are rather weak in all of the four climatic regions. Despite notable differences in the soil moisture–temperature coupling strengths among the four seasons, there are still some features in common. There appears to be hot spots of soil moisture–temperature coupling in the transitional zones between wet and dry regimes in all seasons.

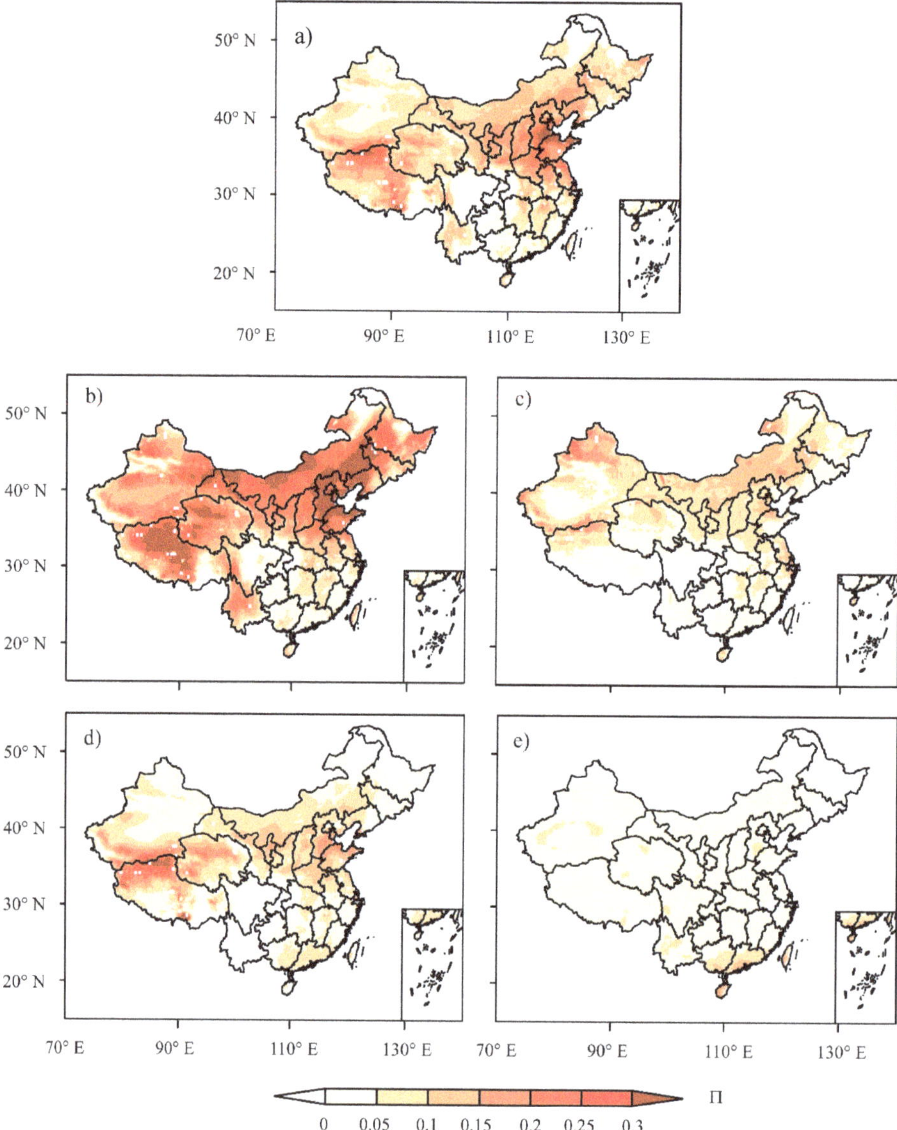

Figure 2. Soil moisture–temperature coupling over China during the period 1980–2013. (**a**) Whole year; (**b**) spring (MAM, from March to May); (**c**) summer (JJA, from June to August); (**d**) autumn (SON, from September to November); (**e**) winter (DJF, from December to February).

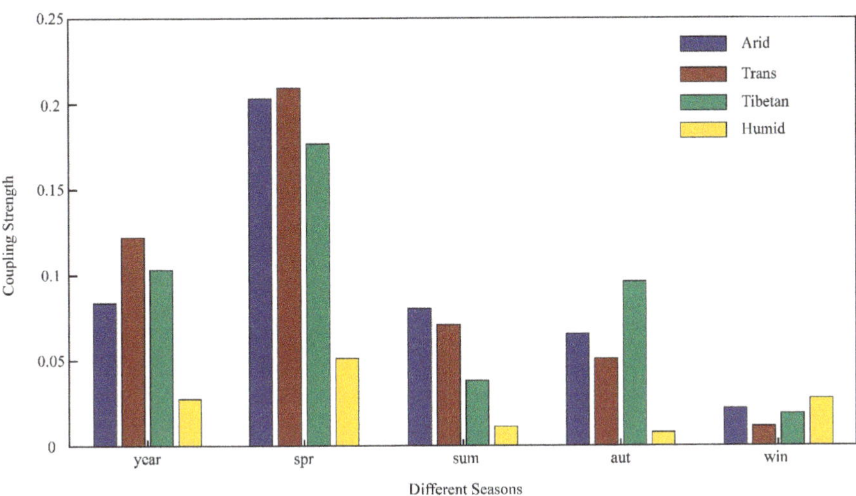

Figure 3. Soil moisture–temperature coupling strengths in different seasons in the typical climate regions over China during the period 1980–2013.

In order to understand how the soil moisture–temperature coupling strengths are linked to the soil moisture amount, Figure 4 shows the density scatter plot of coupling strengths against the soil moisture amount in spring. It appears that, principally, the coupling strengths are linearly related to the soil moisture amounts across China; when soil increases, the strength of soil moisture-temperature coupling decreases, and this linear relationship is particularly clear when the soil moisture amount is more than 0.2 m^3/m^3. When soil is too dry, the coupling is not sensitive to soil moisture amount, and where there is too much soil moisture, especially in the humid regions, the density of data points is also high and shows that the coupling strengths are rather low. This result is also consistent with other studies [11,13].

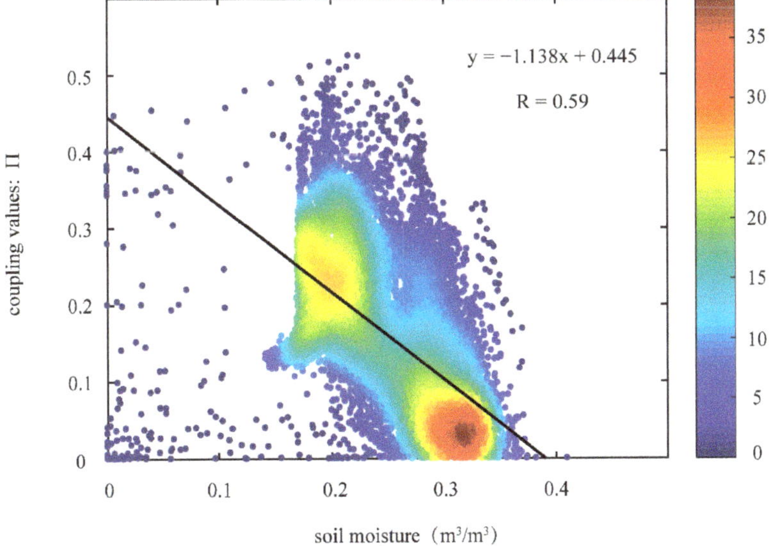

Figure 4. The density scatter plot of coupling strengths against the soil moisture amount in spring.

3.2. Coupling Anomalies in Heatwaves

As seen in Figure 2, the soil moisture–temperature coupling is strongest in spring than in other seasons in China; however, many studies have reported heat wave amplifications through the feedback loops between soil moisture deficit and temperature in summer [18,44]. To understand the detailed processes of soil moisture–temperature coupling in the heatwave events, we conducted two case studies on daily scales.

3.2.1. Case 1: Heatwave of Southeast China in Summer 2013

In the summer of 2013, Southeast China experienced abnormally high temperatures, which broke the heat records for the past 141 years and led to an unprecedented heatwave across China [1]. This unprecedented anomalies reached high values from 23 July to 14 August [3]. This disaster caused about US$10 billion in crop damage, and a total of 5758 Heatwave-related illness cases were reported [40–47].

Figure 5a illustrates the soil moisture–temperature coupling from 1 June to 30 August, when the heatwave occurred. Figure 5b,c shows the soil moisture and temperature anomalies, referring to the multi-year average of 1980–2013. It appears that the summer temperature was strongly coupled to land surface soil moisture in East China, where the heatwave occurred. Meanwhile, there were significant soil moisture deficits and large-scale positive temperature anomalies, reaching roughly 6 °C, particularly in the middle and lower reaches of the Yangtze River basin. Atmospheric circulation anomalies are generally considered to be the main cause of heatwaves in China, e.g., the movement of the Northwest Pacific subtropical high [46,48,49]. However, land surface feedback on the atmosphere have been found to be an important factor for heatwaves over China, which may contribute 30–70% of the high temperature anomalies [22]. Our finding from Figure 5a–c is very likely to tell such a story that soil moisture deficit resulted in, at least partly, a significant heatwave through the coupled processes between land and atmosphere.

To better understand the temporal evolution of the heat wave, we show in Figure 5d–f the changes of the temperature anomalies (T') and the heat anomalies ($H' - H_p'$) with time. Figure 5d shows the positive heat anomalies before the heatwave occurred (12–23 July) in Southern China, which are related to the soil moisture deficit and lead to enhanced evaporative stress. Figure 5e illustrates the mega-heatwave from 24 July to 16 August in Southern China, when both the temperature anomalies (T') and the heat anomalies ($H' - H_p'$) reached their maximum values with the largest spatial coverage. Figure 5f shows the temperature anomalies (T') and the heat anomalies ($H' - H_p'$) during 17–28 August, when the heatwave had almost vanished with largely reduced temperature and heat anomalies, as well as their spatial coverages. Further, in Figure 5g, we show the temporal variations of $\left(H' - H_p'\right)$ and T' from 1 June to 13 August, which are averaged within a small region in the epicenter of the heatwave (marked in Figure 5e). The right Y-axis indicates the metric π, and the left Y-axis indicates the anomalies of $\left(H' - H_p'\right)$ and T'. It clearly shows how soil moisture deficit contributed to the enhancement of heat and temperature anomalies; it is obvious that the land–atmosphere coupling was strongest with the largest π values during the mega heatwave (23 July–18 August).

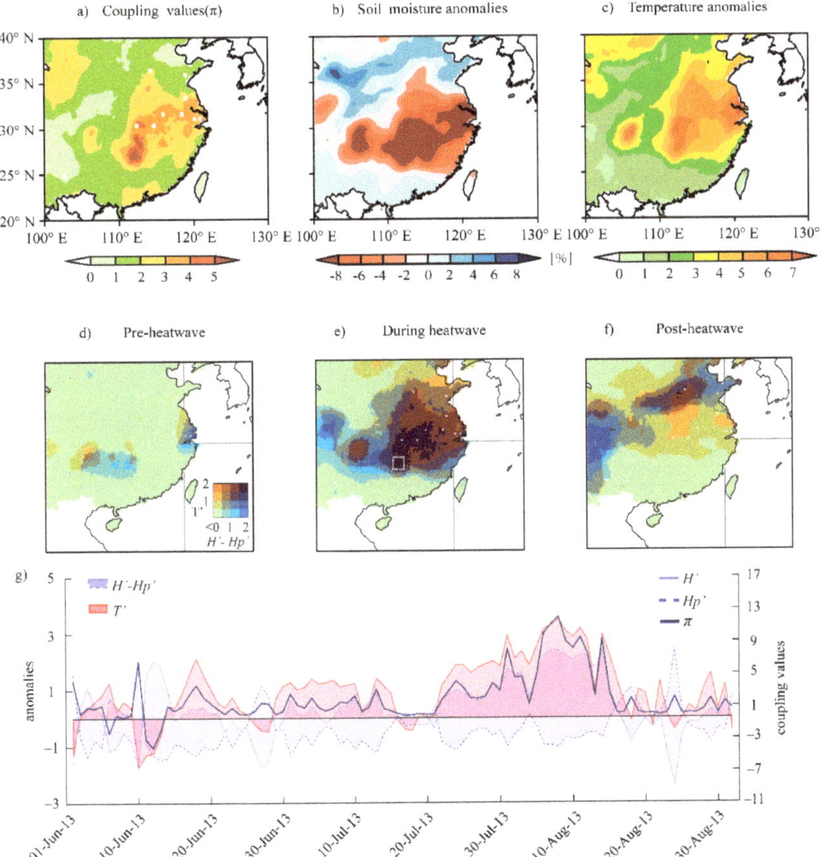

Figure 5. The soil moisture–temperature coupling and related processes during the summer heatwave of Southeast China during 24 July–16 August 2013. (**a**) The coupling metric π; (**b**) the soil moisture anomalies and (**c**) the temperature anomalies, referring to the multi-year average of 1980–2013; (**d**) pre-heatwave (12–23 July); (**e**) the occurrence of the heatwave (24 July–16 August); (**f**) the days after the heatwave (17–28 August); (**g**) daily time series of T', H', H_p', and the coupling metric π.

3.2.2. Case 2: Heatwave of North China in Summer 2009

There was a heatwave from 20 June to 4 July in 2009 in North China. Although it was not as severe as the mega-heatwave in Southeast China in 2013, the continuous hot weather in the North China plain was rare since 1949 [50]. At the end of June, an unprecedented heatwave hit Hebei and Shandong provinces, setting new temperature records. By early July, the heatwave gradually dissipated in North China and the high temperature shifted to Southern China [51]. The cause analysis of this event mainly focused on the effect of El Niño and is typical of the high-pressure systems [49]. In addition, the soil moisture deficit of North China would increase the sensible heat flux and influence the atmospheric boundary layer temperature, conducive to strengthening the subtropical high and causing a heatwave [52]. Figure 6a–c depicts the coupling metric π, as well as the soil moisture and temperature anomalies referring to multi-year average of 1980–2013 during the heatwave (20 June–4 July) in North China in 2009. Where there is the strongest coupling, there is significant soil moisture deficit and maximum temperature anomalies.

Figure 6d–f is analogous to Figure 5d–f, dividing the study period into pre-heatwave (the left panel), the mega heatwave (the middle panel), and post-heatwave (the right panel). It appears that the heat anomalies ($H' - H_p'$) and the temperature anomalies (T') reached their maximum values during the heatwave, while such anomalies existed in neither the left nor the right panel. In Figure 6f, we show $\left(H' - H_p'\right)$ and T' for the period 1 June–31 August, averaged within a small region in the epicenter of the heatwave (marked in Figure 6e). Similarly, it is obvious that the land–atmosphere coupling is strongest with the largest π values during the mega heatwave 20 during June to 4 July, when there is the largest heat and temperature anomalies associated with a significant soil moisture deficit.

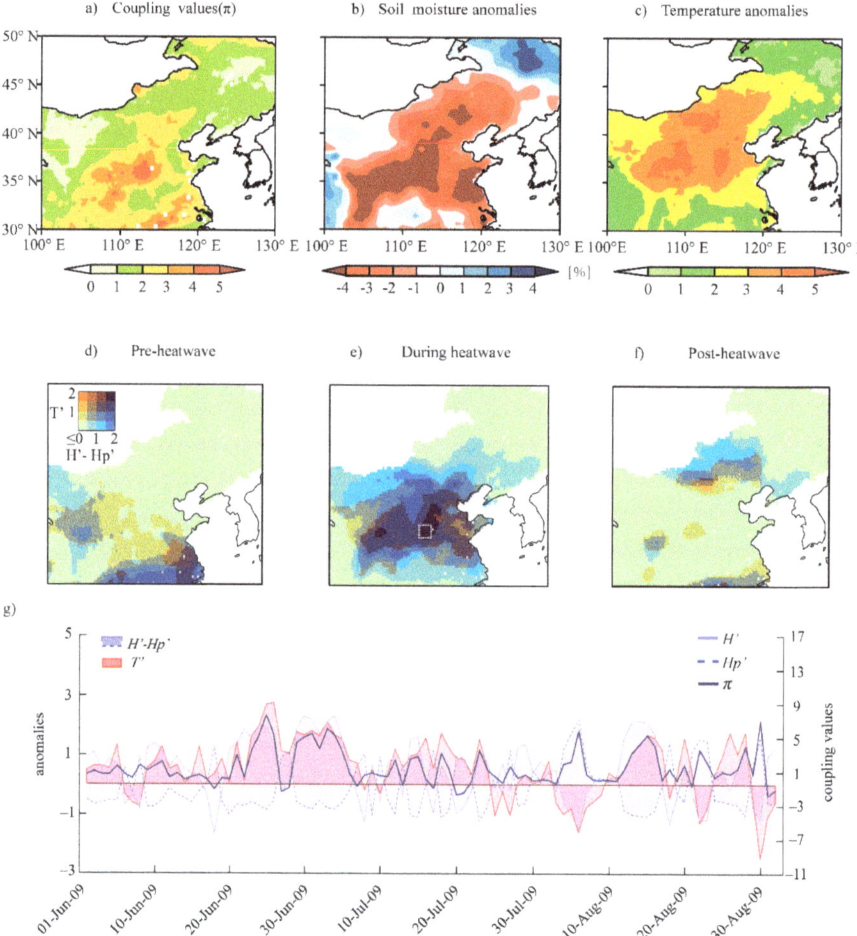

Figure 6. The soil moisture–temperature coupling and related processes during the summer heatwave of North China during 20 June–4 July 2009. (**a**) The coupling metric π; (**b**) the soil moisture anomalies and (**c**) the temperature anomalies referring to the multi-year average of 1980–2013; (**d**) pre-heatwave (12–19 June); (**e**) the occurrence of the heatwave (20 June–5 July); (**f**) the days after the heatwave (6–13 July); (**g**) daily time series of T', H', H_p', and the coupling metric π.

4. Conclusions and Discussion

This study attempted to utilize the GELAM and ERA-Interim datasets to study land–atmosphere coupling in China for the period 1980–2013. The key findings are as follows.

Hot spots of soil moisture–temperature coupling were found in North China and over the Tibetan Plateau, which indicate that the soil moisture–temperature coupling is strongest in the transitional climate zones. These results are in agreement with [9,14]. The seasonality of soil moisture–temperature coupling strength has marked regional variation, which suggests that soil moisture–temperature coupling strength is stronger in Northern China than in the southern, and coupling is relatively stronger in spring, followed by summer and autumn, and insignificant in winter. In spring, soil moisture–temperature coupling is stronger within dry areas, and these regimes are mainly water-limited regions, where evaporation depends on the supply of water.

Case studies involving the 2013 Southeast China heatwave and the 2009 North China heatwave were conducted to understand the role of soil moisture–temperature coupling and the related heating processes during heatwave events. It was found that enhanced heat and temperature anomalies associated with soil moisture deficit, when the soil moisture–temperature coupling intensifies, could result in enhanced evaporative stress and heat anomalies, which finally leads to enhanced soil moisture coupling with temperature [1,50]. However, there is much debate about the exact physical mechanisms of how heatwaves occur and evolve. Mirelles et al. [11] found that the prevailing persistent synoptic patterns led to warm air advection and clear skies, along with a high atmospheric demand, intensified soil desiccation (causing a strong surface sensible heat flux), causing the mega-heatwaves of 2003 and 2010 in Europe. Zhang et al. [53] supports that soil moisture in spring and early summer may be an important contributor to heatwaves in July via positive subtropical high anomalies. Several studies show that the formation mechanism of heatwaves is not only caused by a certain external force, but may be influenced by circulation systems, external forcing, or local effect, and soil moisture deficit may contribute directly and indirectly to all of these processes [10].

Land–atmosphere coupling involves water, energy, and chemical elements, which affect different processes in the hydrological cycle and thus play a critical role in the climate system [8]. However, in-situ observations of soil moisture and land surface fluxes are scarce and uncertain at the appropriate scale, which has caused great difficulties in the study of land–atmosphere coupling. The recent advances in satellite remote sensing have provided near-real-time datasets for us, and reanalysis data can also provide long-term databases for such studies. The long-term memory of soil moisture could help better understand the land–atmosphere interactions and may provide valuable information in weather forecasts to aid the management of extremely warm climates.

Other dynamic relations between soil moisture and atmosphere coupling were ignored in this study, which may also affect the participation of sensible heat and latent heat. Furthermore, the ultimate causal relationship was not demonstrated in the diagnostics between soil moisture and temperature coupling. The physical mechanism and causal relationship between soil moisture and temperature still need to be further explored in future work.

Author Contributions: Q.Y. and G.W. conceived and designed the overall project, and wrote most of sections of the manuscript. C.Z. and D.F.T.H. performed methodology and data analysis. D.L., X.M. and M.Z. supplied suggestions and comments for the manuscript. All authors have read and agreed the published version of the manuscript.

Funding: This study was funded by the National Natural Science Foundation of China (41875094, 41850410492).

Acknowledgments: All authors are grateful to the anonymous reviewers and editors for their constructive comments on earlier versions of the manuscript.

Conflicts of Interest: The authors declare no conflict of interest.

References

1. Sun, Y.; Zhang, X.; Zwiers, F.W.; Song, L.C.; Wang, H.; Yin, H.; Ren, G.Y. Rapid increase in the risk of extreme summer heat in Eastern China. *Nat. Clim. Chang.* **2014**, *4*, 1082–1085. [CrossRef]
2. Zhu, Z. An Analysis of Drought Event and Its Causation in Jianghuai Region During Summer 2013. *J. North China Inst. Aerosp. Eng.* **2016**, *26*, 37–40.

3. Gong, Z.Q.; Wang, Y.J.; Wang, Z.Y.; Ma, L.J.; Sun, C.H.; Zhang, S.Q. Briefly Analysis on Climate Anomalies and Causations in Summer 2013. *Meteorol. Mon.* **2014**, *40*, 119–125.
4. Xia, Y.; Xu, H.M. Circulation characteristics and causes of the summer extreme high temperature event in the middle and lower reaches of the Yangtze River of 2013. *J. Meteorol. Sci.* **2017**, *37*, 60–69.
5. Jaeger, E.B.; Seneviratne, S.I. Impact of soil moisture-atmosphere coupling on European climate extremes and trends in a regional climate model. *Clim. Dyn.* **2011**, *36*, 1919–1939. [CrossRef]
6. Dong, J.Z.; Crow, W.T. The Added Value of Assimilating Remotely Sensed Soil Moisture for Estimating Summertime Soil Moisture-Air Temperature Coupling Strength. *Water Resour. Res.* **2018**, *54*, 6072–6084. [CrossRef]
7. Gevaert, A.I.; Miralles, D.G.; Jeu, R.A.M.; Schellekens, J.; Dolmon, A.J. Soil Moisture-Temperature Coupling in a Set of Land Surface Models. *J. Geophys. Res. Atmos.* **2017**, *123*, 1481–1498. [CrossRef]
8. Seneviratne, S.I.; Corti, T.; Davin, E.L.; Hirschi, M.; Jaeger, E.B.; Lehner, I.; Orlowsky, B.; Teuling, A.J. Investigating soil moisture-climate interactions in a changing climate: A review. *Earth-Sci. Rev.* **2010**, *99*, 125–161. [CrossRef]
9. Koster, R.D.; Dirmeyer, P.A.; Guo, Z.; Bonan, G.; Chan, E.; Cox, P.; Gordon, C.T.; Kanae, S.; Kowalczyk, E.; Lawrence, D.; et al. Regions of strong coupling between soil moisture and precipitation. *Science* **2004**, *305*, 1138–1140. [CrossRef]
10. Fischer, E.M.; Seneviratne, S.I.; Luthi, D.; Schar, C. Contribution of land-atmosphere coupling to recent European summer heat waves. *Geophys. Res. Lett.* **2007**, *34*, L06707. [CrossRef]
11. Miralles, D.G.; Teuling, A.J.; Heerwaarden, C.C.; Arellano, J.V. Mega-heatwave temperatures due to combined soil desiccation and atmospheric heat accumulation. *Nat. Geosci.* **2014**, *7*, 345–349. [CrossRef]
12. Schwingshackl, C.; Hirschi, M.; Seneviratne, S.I. Quantifying spatiotemporal variations of soil moisture control on surface energy balance and near-surface air temperature. *J. Clim.* **2017**, *30*, 7105–7124. [CrossRef]
13. Koster, R.D.; Schubert, S.D.; Suarez, M.J. Analyzing the concurrence of meteorological droughts and warm periods, with implications for the determination of evaporative regime. *J. Clim.* **2009**, *22*, 3331–3341. [CrossRef]
14. Seneviratne, S.I.; Luthi, D.; Litschi, M.; Schar, C. Land-atmosphere coupling and climate change in Europe. *Nature* **2006**, *443*, 205–209. [CrossRef]
15. Vidale, P.L.; Luthi, D.; Wegmann, R.; Schar, C. European summer climate variability in a heterogeneous multi-model ensemble. *Clim. Chang.* **2007**, *81*, 209–232. [CrossRef]
16. Koster, R.D.; Guo, Z.C.; Dirmeyer, P.A.; Bonan, G.; Chan, E.; Cox, P.; Davies, H.; Gordom, C.T.; Kowalczyk, E.; Lawrence, D.; et al. GLACE: The Global Land—Atmosphere Coupling Experiment. Part I: Overview. *J. Hydrometeorol.* **2006**, *7*, 611–625. [CrossRef]
17. Miralles, D.G.; Berg, M.J.; Teuling, A.J.; Jeu, R.A.M. Soil moisture-temperature coupling: A multiscale observational analysis. *Geophys. Res. Lett.* **2012**, *39*, 2–7. [CrossRef]
18. Seneviratne, S.I.; Wilhelm, M.; Stanelle, T.; Hurk, B.; Hagemann, S.; Berg, A.; Cheruy, F.; Higgins, M.E.; Meier, A.; Brovkin, V.; et al. Impact of soil moisture-climate feedbacks on CMIP5 projections: First results from the GLACE-CMIP5 experiment. *Geophys. Res. Lett.* **2013**, *40*, 5212–5217. [CrossRef]
19. Hirsch, A.L.; Pitman, A.J.; Seneviratne, S.I.; Evans, J.P.; Haverd, V. Summertime maximum and minimum temperature coupling asymmetry over Australia determined using WRF. *Geophys. Res. Lett.* **2014**, *41*, 1546–1552. [CrossRef]
20. Whan, K.; Zscheischler, J.; Orth, R.; Shongwe, M.; Rahimi, M.; Asare, E.O.; Seneviratne, S.I. Impact of soil moisture on extreme maximum temperatures in Europe. *Weather Clim. Extrem.* **2015**, *9*, 57–67. [CrossRef]
21. Liu, D.; Wang, G.L.; Mei, R.; Yu, Z.B.; Gu, H.H. Diagnosing the Strength of Land–Atmosphere Coupling at Subseasonal to Seasonal Time Scales in Asia. *J. Hydrometeorol.* **2013**, *15*, 320–339. [CrossRef]
22. Zhang, J.Y.; Wu, L.Y.; Dong, W.J. Land-atmosphere coupling and summer climate variability over East Asia. *J. Geophys. Res. Atmos.* **2011**, *116*, 1–14. [CrossRef]
23. Liu, D.; Yu, Z.B.; Zhang, J.Y. Diagnosing the strength of soil temperature in the land atmosphere interactions over Asia based on RegCM4 model. *Glob. Planet. Chang.* **2015**, *130*, 7–21. [CrossRef]
24. Gallego-Elvira, B.; Taylor, C.M.; Harris, P.P.; Ghent, D.; Veal, K.L.; Folwell, S.S. Global observational diagnosis of soil moisture control on the land surface energy balance. *Geophys. Res. Lett.* **2016**, *43*, 2623–2631. [CrossRef]
25. Dirmeyer, P.A. The terrestrial segment of soil moisture-climate coupling. *Geophys. Res. Lett.* **2011**, *38*, 116702. [CrossRef]

26. Teuling, A.J.; Seneviratne, S.I.; Stockli, R.; Reichstein, M.; Moors, E.; Ciais, P.; Luyssaert, S.; Hurk, B.; Ammann, C.; Bernhofer, C.; et al. Contrasting response of European forest and grassland energy exchange to heatwaves. *Nat. Geosci.* **2010**, *3*, 722–727. [CrossRef]

27. Balsamo, G.; Agusti-Parareda, A.; Albergel, C.; Arduini, G.; Beljaars, A.; Bidlot, J.; Blyth, E.; Bousserez, N.; Boussetta, S.; Brown, A.; et al. Satellite and in situ observations for advancing global Earth surface modelling: A Review. *Remote Sens.* **2018**, *10*, 2038. [CrossRef]

28. Levine, P.A.; Randerson, J.T.; Swenson, S.C.; Lawrence, D.M. Evaluating the strength of the land-atmosphere moisture feedback in Earth system models using satellite observations. *Hydrol. Earth Syst. Sci.* **2016**, *20*, 4837–4856. [CrossRef]

29. Diffenbaugh, N.S.; Pal, J.S.; Giorgi, F.; Gao, X.J. Heat stress intensification in the Mediterranean climate change hotspot. *Geophys. Res. Lett.* **2007**, *34*. [CrossRef]

30. Gao, X.; Shi, Y.; Song, R.; Giorgi, F.; Wang, Y.; Zhang, D. Reduction of future monsoon precipitation over China: Comparison between a high resolution RCM simulation and the driving GCM. *Meteorol. Atmos. Phys.* **2008**, *100*, 73–86. [CrossRef]

31. Liang, Y.L.; Han, M.C.; Bai, L.; Li, M.H. Spatial-temporal distribution and variation characteristics of the agricultural climate resources over recent 30 years in China. *Agric. Res. Arid Areas* **2015**, *33*, 259–267.

32. Maó, F.; Tang, S.H.; Sun, H.; Zhang, J.H. A Study of Dynamic Change of Dry and Wet Climate Regions in the Tibetan Plateau over the Last 46 Years. *Chin. J. Atmos. Sci.* **2008**, *32*, 499–507.

33. Fan, K.K.; Zhang, Q.; Sun, P.; Song, C.Q.; Yu, H.Q.; Zhu, X.D.; Shen, Z.X. Effect of soil moisture variation on land surface air temperature over Tibetan Plateau. *Acta. Geogr. Sin.* **2019**, *74*, 1–16.

34. Cheng, G.D.; Jin, H.J. Groundwater in the permafrost regions on the qinghai-tibet plateau and it changes. *Hydrogeol. J.* **2013**, *21*, 5–23. [CrossRef]

35. Miralles, D.G.; Holmes, T.R.H.; Jeu, R.A.M.D.; Gash, J.H.; Meesters, A.G.C.A.; Dolman, A.J. Global land-surface evaporation estimated from satellite-based observations. *Hydrol. Earth Syst. Sci.* **2010**, *15*, 453–469. [CrossRef]

36. Priestley, C.H.B.; Taylor, R.J. On the assessment of surface heat flux and evaporation using large-scale parameters. *Weather Rev.* **1972**, *100*, 81–92. [CrossRef]

37. Dee, D.P.; Uppala, S.M.; Simmons, A.J.; Berrisford, P.; Poli, P.; Kobayashi, S.; Andrae, U.; Balmaseda, M.A.; Balsamo, G.; Bauer, P.; et al. The ERA-Interim reanalysis: Configuration and performance of the data assimilation system. *Q. J. R. Meteorol. Soc.* **2011**, *137*, 553–597. [CrossRef]

38. Penman, H.L. Natural evaporation from open water, bare and grass. *Proc. R. Soc. A* **1948**, *193*, 120–145.

39. Zhang, Q.; Xiao, M.Z.; Singh, V.P.; Liu, L.; Xu, C.Y. Observational evidence of summer precipitation deficit-temperature coupling in China. *J. Geophys. Res. Atmos.* **2015**, *120*, 10040–10049. [CrossRef]

40. Mueller, B.; Seneviratne, S.I. Hot days induced by precipitation deficits at the global scale. *Proc. Natl. Acad. Sci. USA* **2012**, *109*, 12398–12403. [CrossRef]

41. Li, M.M.; Ma, Z.; Gu, H.; Yang, Q.; Zhang, Z. Production of a combined land surface data set and its use to assess land-atmosphere coupling in China. *J. Geophys. Res. Atmos.* **2017**, *122*, 948–965. [CrossRef]

42. Shi, X.; Wang, Y.; Xu, X.; Shi, X.; Wang, Y.; Xu, X. Effect of mesoscale topography over the Tibetan Plateau on summer precipitation in China: A regional model study. *Geophys. Res. Lett.* **2008**, *35*. [CrossRef]

43. Wang, Z.; Yang, S.; Duan, A.; Hua, W.; Ullah, K.; Liu, S. Tibetan Plateau heating as a driver of monsoon rainfall variability in Pakistan. *Clim. Dyn.* **2018**, *52*, 6121–6130. [CrossRef]

44. Vogel, M.M.; Orth, R.; Cheruy, F.; Hagemann, S.; Lorenz, R.; Hurk, B.J.J.M.; Seneviratne, S.I. Regional amplification of projected changes in extreme temperatures strongly controlled by soil moisture-temperature feedbacks. *Geophys. Res. Lett.* **2017**, *44*, 1511–1519. [CrossRef]

45. LeComte, D. International Weather Highlights 2013: Super Typhoon Haiyan, Super Heat in Australia and China, a Long Winter in Europe. *Weatherwise* **2014**, *67*, 20–27. [CrossRef]

46. Luo, M.; Lau, N.C. Heat waves in southern China: Synoptic behavior, long-term change, and urbanization effects. *J. Clim.* **2017**, *30*, 703–720. [CrossRef]

47. Gu, S.H.; Huang, C.R.; Bai, L.; Chu, C.; Liu, Q.Y. Heat-related illness in China, summer of 2013. *Int. J. Biometeorol.* **2016**, *60*, 131–137. [CrossRef]

48. Sun, J.; Wang, H.; Yuan, W. Decadal Variability of the Extreme Hot Event in China and Its Association with Atmospheric Circulations. *Clim. Environ. Res.* **2011**, *2*, 199–208.

49. Qian, W.H.; Ding, T. Atmospheric anomaly structures and stability associated with heat wave events in China. *Chin. J. Geophys.* **2012**, *55*, 1487–1500.
50. Zhang, Y.X.; Zhang, S.J. Causation Analysis on a Large –Scale Continuous High Temperature Process Occurring in North China Plain. *Meteorol. Mon.* **2010**, *36*, 8–13.
51. Li, Y.; Ma, B.S.; Yang, X.; Zhang, J.Y. Characteristics of summer heat waves in China Mainland and the relationship between Eastern-/Central-Pacific EI Nino and heat wave events. *J. Lanzhou Univ. Nat. Sci.* **2018**, *54*, 711–721.
52. Wang, L.W.; Zhang, J. Relationship Between summer heat waves and soil moisture in North-East China. *J. Meteorol. Sci.* **2015**, *35*, 558–564.
53. Zhang, W.L.; Zhang, J.Y.; Fan, G.Z. Dominant modes of dry- and wet-season precipitation in southwestern China. *Chin. J. Atmos. Sci.* **2014**, *38*, 590–602.

MDPI
St. Alban-Anlage 66
4052 Basel
Switzerland
Tel. +41 61 683 77 34
Fax +41 61 302 89 18
www.mdpi.com

Atmosphere Editorial Office
E-mail: atmosphere@mdpi.com
www.mdpi.com/journal/atmosphere